This book is to be returned on or before

**Books are to be returned on or before
the last date below.**

06 MAR 1996

LIBREX—

IEE DIGITAL ELECTRONICS AND COMPUTING SERIES 1

SERIES EDITORS: S. L. HURST AND M. W. SAGE

STOCHASTIC AND DETERMINISTIC AVERAGING PROCESSORS

STOCHASTIC AND DETERMINISTIC AVERAGING PROCESSORS

P. MARS, M.Sc., Ph.D., C. Eng., M.I.E.E., Sen. Mem. I.E.E.E.
Head of the School of Electronic & Electrical Engineering,
Robert Gordon's Institute of Technology,
Aberdeen AB9 1FR
Scotland
and
W.J. POPPELBAUM, M.S., Ph.D., F.I.E.E.E.
Director
The Information Engineering Laboratory,
Department of Computer Science
University of Illinois
USA

PETER PEREGRINUS LTD.
On behalf of the
Institution of Electrical Engineers

Published by: The Institution of Electrical Engineers, London
and New York
Peter Peregrinus Ltd., Stevenage, UK, and New York

British Library Cataloguing in Publication Data

Mars, P.
  Stochastic and deterministic
  averaging processors.
  1.   Mathematical models
  2.   Stochastic processes
  I.   Title   II.   Poppelbaum, W. J.
  519.2   QA402

ISBN: 0 906048 44 3

621. 3804' 3
MAR

Typeset at the Alden Press Oxford London and Northampton
Printed in England by A. Wheaton & Co., Ltd., Exeter

Contents

Preface

This book is concerned with the design and application of processors which use an information representation such that *averaging* must be employed to extract useful information. In particular we consider mainly time averaging of information signals consisting of pulses in predetermined time slots, i.e. serial synchronous digital systems. It should be noted that there do exist averaging systems in which space-averages replace time-averages, and time-average systems which are essentially asynchronous.

Under the above restrictions we can still choose between *weighted* (binary) *systems* and *unweighted systems*. In the latter category we can distinguish at least two possible approaches:

(1) 'Open-ended' unweighted sequences of randomly occuring pulses: this is *time stochastic processing* (t.s.p.), in which the probability of occurence of a pulse is the information carrier.

(2) Fixed length blocks, representing numbers in an m out of n system in some (usually deterministic) manner: this is *burst processing* (b.p.), in which the average over many blocks carries the information.

It is quite evident that neither t.s.p or b.p. uses optimal ways of encoding information as far as bandwidth is concerned. We will show subsequently that for 1% precision one needs roughly 10^4 slots in t.sp. and 10^2 slots in b.p. For 10% precision things look better $-$ 10^2 slots in t.s.p. and 10 slot in b.p. are now sufficient. However, in 10 slots we could have transmitted 1024 different binary values $-$ even b.p. needs therefore roughly 100 times more bandwidth than binary computation. But now comes the important point: in many applications we underutilise the bandwidth of a dedicated channel anyway. Typical examples are:

(1) All cases in which a channel of standardised configuration is used to transmit ON/OFF type signals (e.g. engine-room telegraph).

(2) Transmission of telemetry signals corresponding to slowly varying variables of low precision (e.g. antenna position indicator).

(3) Systems in which the computational speed of the processors is the limiting factor. It is clearly useless to design a 10 MHz, 10 bit precision pulse code modulation (p.c.m.) system to feed a 10 or $100\,\mu$s multiplication circuit!

(4) Systems in which the speed limitations of the output activators form the bandwidth bottleneck (e.g. rudder and aileron controls).

Ignoring for the moment the issue of 'wasted bandwidth' we should consider the significant advantages of t.s.p and b.p. Examples are:

(1) *Simple computational circuits:* arithmetic in t.s.p. is extremely simple: AND gates multiply and trivial combinational logic may be used for addition. In b.p., arithmetic is also quite simple: counters or current summing registers are the essential elements.

(2) *Error tolerance:* both t.s.p. and b.p. use an unweighted representation and averaging: an occasional supernumeracy or missing pulse does not matter.

(3) *Constant availability of results:* t.s.p. with integrators and b.p. with summation in a length n register gives a useful output at all times; even short observations can be acted upon.

(4) *Use for numerical and communication applications:* t.s.p. and especially b.p., can cope equally well with data transmission and voice (or video) signals.

(5) *Loose clocking and synchronisation:* both t.s.p. and b.p. interpret averages for processing – this usually eliminates clocking and initialisation signal difficulties.

(6) *Reliability:* the considerable simplicity of circuit realisation leads to higher reliability. Also checking using local coding techniques is usually affordable.

(7) *Ease of multiplexing:* switching noise is eliminated and reassembly of frames (i.e. blocks in b.p.) is easy, as in all p.c.m. systems.

Essentially, t.s.p. and b.p. are forms of p.c.m. and as such partake of the well known p.c.m. advantage: easy signal renormalisation in long-distance transmission (whether a distance is 'long' is determined by the noise of the environment). The systems we describe are doubly noise-proof; pulses are usually unaltered, but if a few are altered we can live with the result.

In Chapter 1 we review the basic types of averaging processors, and Chapter 2 outlines the fundamentals of t.s.p. with particular emphasis on the critical problems of random number generation and output interface design. The original development of t.s.p. took place independently at the University of Illinois and STL in England during 1965. Since that time, stochastic processing has become a subject of research in many establishments throughout the world. The First International Symposium on Stochastic Computing held in Toulouse at the end of 1978, gave a good indication of the steady progress which has been achieved. Chapter 3 describes some of the significant practical applications of t.s.p.

A recent development has been the demonstration that t.s.p. provides an optimum synthesis technique for stochastic learning automata. This is an exciting research area and a classic application of t.s.p. in that direct processing of random sequences is involved. In Chapter 4 we consider the fundamental types and basic synthesis techniques for stochastic learning automata and describe some important applications. Chapter 5 concludes the book with a detailed treatment of b.p. encoding, arithmetic techniques and an outline of significant potential applications.

P.M.
W.J.P.
May 1980

Acknowledgements

Both authors wish to record the benefit of numerous discussions with Prof. Brian Gaines. We owe a particular debt to our research students on both sides of the Atlantic. Their numerous significant contributions over the years are self evident from the references cited. We acknowledge the assistance of Frank Serio at Illinois and the financial support of various organisations; in particular the SRC in the UK, and, in the USA, the ONR and AEC.

Special thanks to Yvonne Mars for demonstrating infinite patience in typing the entire manuscript.

Figs. 1.5 to 1.21 and 3.1 to 3.10 are reproduced by kind permission of Academic Press. In addition, Figs. 2.10 to 2.13 are by permission of the IEEE and Figs. 2.20 to 2.24 and 3.13 to 3.16 are reproduced by courtesy of the IEE.

Finally, thanks to Bertie McK. Davidson of RGIT for drawing most of the diagrams.

Chapter 1

Basic types of averaging processors

The field of computation and communication, initially purely analogue, has been almost entirely taken over by weighted binary digital techniques. This is simply because the circuit parameter tolerances are so much greater when one makes a binary decision on each pulse; also the base two representation is relatively efficient. The advent of microcomputers and packet switching only reinforced our belief in these binary digital implementations and it may seem preposterous to even suggest alternative, albeit digital, technologies. What we mean by 'alternative' is 'nonweighted binary'. Historically, nonweighted binary systems have played an important role. For example, the earliest digital computers were digital differential analysers (d.d.a.s) in which one simply counted the number of pulses in a given time or in a given number of time slots. A digital differential analyser therefore uses a unary number representation. Each pulse is a marker and one counts markers. Sometimes, of course, the count will have to be displayed, one then uses a binary, octal, decimal or hexadecimal representation. When the pulses are used in a control system (actuators for rudders, antennas etc.) it is usually not necessary to specifically refer to a number system.

More recently, considerable success has been obtained by *stochastic* or *probablistic unary encoding*.[1-7] Here a number is represented by the probability of appearance of a pulse in a given time slot. The ultrasimplicity of the computational circuits in this stochastic method is well known, as is the failsoft behaviour. The crux of the method is to assess the probabilities by measuring the average frequency (the probability being the limit for an infinite number of time slots). Unfortunately, as we shall show, the number of time slots to obtain 10% midrange-precision (with 68% likelihood) is 10^2, for 1% we need 10^4, for 0·1%, 10^6, i.e. the precision increases only as the square root of the number of time slots. Furthermore, one must make quite sure that the random pulse sequences which represent variables to be processed are uncorrelated. In practice there are many applications (e.g. stochastic feedback controls) in which 1% accuracy is quite sufficient. With a 10 MHz clock this means a computation time of 10^{-3}s, an often acceptable value. A recent important result is that time stochastic processing provides an optimum synthesis technique for stochastic learning automata.[93-114]

In addition to time stochastic processors we consider both *bundle* and *ergodic processors*. Bundle processors essentially map the time slots in an infinite stochastic sequence onto the wires of a bundle.[8] Although considerably more complex than time stochastic processors and of more limited accuracy (finite number of wires), bundle processing has two outstanding properties:

(*a*) immediate availability of all results

(*b*) excellent failsoft behaviour when certain special representations of numbers are chosen.

There is no doubt that bundle processing is the optimal case of diffuse computers. No portion of the hardware is in itself of any importance, and as long as the statistical properties of the system as a whole remain unchanged, the results can be trusted. It is superfluous to point out that such diffuse (or 'distributed') computers will be of the highest importance in performing critical functions under emergency conditions. In ergodic processing each wire of a bundle carries a stochastic sequence whose time average is equal to the instantaneous cross-sectional average of the bundle.[9] Such processors possess the advantages of bundle machines with additional checking capability provided by time stochastic processing.

Another form of average processing to be considered known as *burst processing*,[10, 11] is the result of attempting to extend averaging techniques to a deterministic unary number representation in such a manner that the complexity of the circuitry would be quite a bit lower than in weighted binary; although not as low as stochastic processing methods. The actual representation uses a unary pulse code modulation frame, e.g. 7 corresponds to a 'burst' consisting of the (first) seven slots of a 10-slot frame being filled. This has the advantage that, as long as the number represented does not change, it is immaterial which 10 adjacent slots ('window') we look at. A window contains always the same information. Should the information change (e.g. from 2 to 8 occupied slots), the contents of the window will increase steadily from the initial to the final value. In practice it is useful to introduce so called block sum registers which look at 10 adjacent slots and output a voltage proportional to the number of 'ones' in the block sum register.

Besides simplicity, burst processing has some additional advantages. There are no correlation difficulties as in stochastics, nor do we usually have to staticize the information as in a microprocessor. Also the burst format is only $2\frac{1}{2}$ times less efficient in bandwidth requirements than weighted binary. It can be used directly for audio and video transmissions. The lost bandwidth is compensated for by good noise-immunity under very high noise conditions (i.e. as much as 10% error rate). Lastly, the accuracy of burst processing increases linearly with the number of slots, as contrasted with the square-root law of stochastics.

1.1 Time stochastic processing

The fundamental idea of time stochastic processing is to use probabilities as information carriers. Now the probability p is defined experimentally by considering the frequency of occurrence of an event (pulse in a time slot) in the limiting case of an infinite number of time slots.* If there are n pulses in N slots for a given wire, and if n/N tends towards a limit as $N \to \infty$, we set

$$p = \underset{N \to \infty}{\text{Lim}} \frac{n}{N}$$

This means, of course, that for a small number of time slots we may obtain an erroneous assessment of the probability and the number which it represents. In Fig. 1.1 we show for a synchronous random pulse sequence (i.e. a sequence with identical pulses which, if they occur, occur with a standard shape within regularly clocked time slots) what can happen. At the top of the figure we see 3 pulses in 10 time slots, leading to the conclusion that the number transmitted is 0·3. In the middle we have an alternative arrangement (equally likely), leading again to 0·3. If now we just take the first 3 slots of the middle sequence (as shown at the bottom of the figure) we would assume that we are trying to transmit 0·66, i.e. 2/3. The important conclusion is that *short sequences are untrustworthy*.

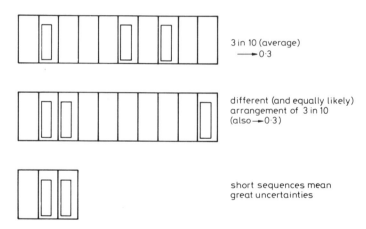

3 in 10 (average)
⟶ 0·3

different (and equally likely) arrangement of 3 in 10 (also ⟶ 0·3)

short sequences mean great uncertainties

Fig. 1.1.   *Uncertainty and sequence length*

The fundamental basis of stochastic processing is to use the theorems of probability theory to perform arithmetic. In Fig. 1.2 this is shown for multiplication. The top sequence input to the AND gate corresponds to probability 0·5, the bottom sequence to 0·4. The probability of an output pulse is the probability of having both incoming time slots occupied simultaneously. If there is no causal relationship between the two sequences (one could think of something as ridiculous as cutting out every 5th pulse of the top sequence to obtain the bottom sequence), i.e. if they are uncorrelated, the probability of simultaneous occupancy is the product

* Readers without the necessary background in basic probability theory should consult a standard introductory text; for example, FELLER, W.: 'An introduction to probability theory and its applications – Vol. 1' (Wiley, 1968).

of the two input probabilities, i.e. 0·2. We have cheated ever so slightly in Fig. 1.2 by arranging the pulses in such a way that the assessment of probabilities can be made on a small number of time slots. In real life it could happen, although rarely, that we would have to look at thousands of pulses before statistics give the correct result.

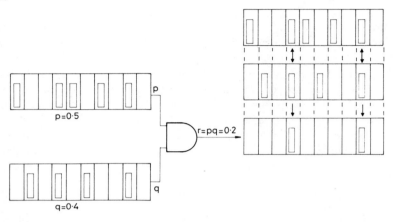

Fig. 1.2   *Time stochastic multiplier*

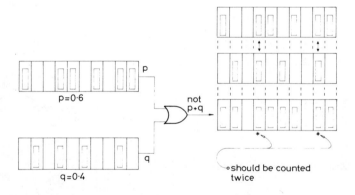

Fig. 1.3.   *Unsound adder*

Encouraged by multiplication we might now turn to the design of Fig. 1.3 for addition. Something should warn us, that the OR-circuit will not work. Here the arithmetic sum of p and q should be unity, i.e. all slots of the output would have to be occupied. Only in the extremely rare case of the two input sequences 'meshing' would this actually be possible. Furthermore, the sum of p and q for $p = 0.8$ and $q = 0.9$ would have to give 1·7, a number no longer representable by a probability and we have overflow. Fig. 1.4 shows a way out of the dilemma. We first scale the

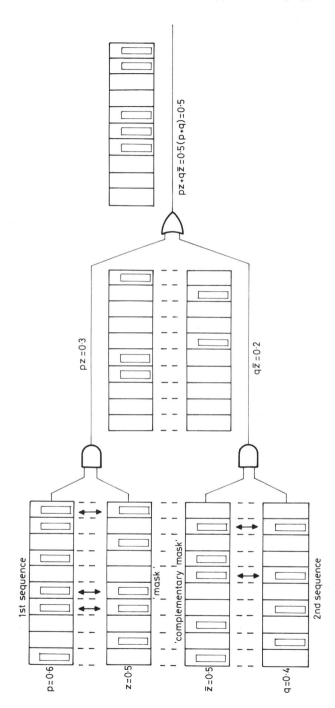

Fig. 1.4 *Workable adder*

input sequences by a factor 0·5, i.e. we multiply by a 'masking sequence' of probability 0·5. The masking sequences, although both representing 0·5, are complementary, i.e. where one has pulses, the other one does not, and vice versa. This means that the products cannot have any overlapping pulses, and can be piped into an OR for summation. Note that our scaling also eliminates the overflow problem mentioned above. Instead of masking sequences we can also use two phases of a clock system.

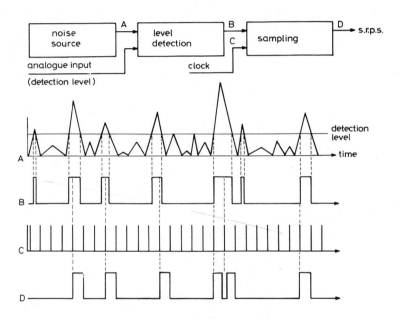

Fig. 1.5.   *Generation of standardised pulses in clocked time slots*
           s.r.p.s. = synchronous random pulse sequence

One last question must be raised in this overview, namely how one can control the number of pulses in a clocked sequence of time slots while conserving randomness. Fig. 1.5 shows the principle. We detect the number of occurrences of noise spikes (Sequence A) higher than an (adjustable) detection level, and normalise the 'above threshold' part of the noise in height (Sequence B). Then we sample by a regular clock (Sequence C), and place a pulse (standardised in height and length) into the next slot if, and only if, the sampling device has detected a pulse in the preceding slot. The resultant Sequence D can be shown to be completely random, but its frequency is controlled by the detection level.[12]

A more rigorous treatment of all operations of arithmetic and the problems of random number generation and output interface design are given in Chapter 2.

1.2   Bundle and ergodic processing

Bundle processing maps the time slots of time stochastic processing onto the wires of a bundle. Now the probability of being 'energised' (i.e. at a 'one-level') at a given instant, for any wire in the bundle, is the carrier of numerical information (see Fig. 1.6). Although most of the thoughts on time stochastic processing can be carried over into bundle processing (e.g. multiplication can be obtained by ANDing pairs of wires of the incoming bundles), there is a fundamental difference. Bundles contain a finite number of wires (perhaps 100 to 1000) while stochastic sequences have theoretically infinitely many slots. The accuracy of a bundle is therefore limited once and for all. But while 100 slots gave about 10% accuracy, 100 wires give obviously 1% accuracy (referred to the maximum number). Thus things are therefore not quite as bad as one might fear. Furthermore, bundles are intrinsically fail-soft (one wire more or less does not matter), can be processed in a distributed manner and furnish a result immediately, i.e. without waiting for some averaging process. Recently optical multichannel designs have made bundle processing economically feasible.[84]

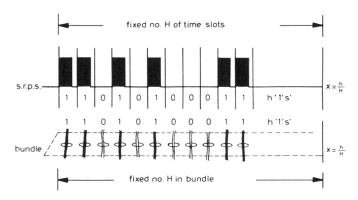

Fig. 1.6.   *Stochastic and bundle processing*

In order to be able to draw bundle processors efficiently, the notation of Fig. 1.7 is used. Visibly the logic (or other) function performed on the pairs of incoming wires (one wire in each pair from each one of the bundles) is summarised by a double-circle around the logic symbol, while the bundles are indicated by very heavy lines. Fig. 1.8 shows some possible bundle operations. We can invert, AND, OR, merge, halve, etc.; the results are not always useful. Note that the halving operation is necessary after merging because the merged bundle has twice as many wires as the original ones. If we cut the merged bundle in half in a random fashion (e.g. 'sawing through the 'upper' half), a random distribution of 'ones' will stay random in the remaining half.

Just as in time stochastic processing, the introduction of negative numbers leads to remapping, i.e. the bundle's cross-sectional average x $(0 \leqslant x \leqslant 1)$ is made a linear function of the number y $(-1 \leqslant y \leqslant +1)$ to be represented. Of course, a signed absolute value system is possible, but certainly not attractive, for the same reasons for which such a system is shunned in conventional digital computers. Fig. 1.9 shows some operations after the remapping $y = 2x - 1$ $(y = 1 - 2x$ would be just as acceptable). Note that multiplication now necessitates a more complicated method than just ANDin all pairs. We have to merge and halve several times and the result is scaled by $1/8$.

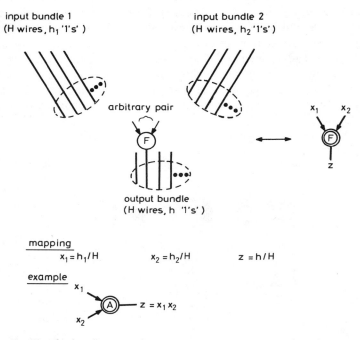

Fig. 1.7.   *Notation for bundle processing*

Division brings up new problems. The so-called 'two-wire system' of time stochastic processing inspires us to use ratio bundles, i.e. to interpret the number to be represented as the (remapped) denominator, divided by the (remapped) numerator. Fig. 1.10 shows some examples of straight, remapped and ratio representations. The remapping used is actually the alternate $y = 1 - 2x$ (with $y = a$ and $x = a'$) mentioned above.

The ratio bundle method leads to a most attractive idea when we add the additional hypothesis that numerator-bundles and denominator-bundles are equally vulnerable. This hypothesis can be satisfied if we assume that each numerator wire is paired-off with a denominator wire, perhaps by twisting them together, so that damage to one will lead to damage to the other. It is then clear that we have not only a *failsoft* system (i.e. tolerant of a few broken wires), but a nearly *failsafe* system (i.e. tolerant of a great number of broken wires). Fig. 1.11 shows this idea, without actually using remapping techniques for numerator and denominator. Fig. 1.12 shows, in the bundle-notation discussed above, how two numbers can be added by distributed circuitry. Any wire or any semiconductor is nonessential, and as long as a reasonable number of transistors, wires, etc., remain intact, the output can be trusted. We have here an example of a truly diffuse computational system.

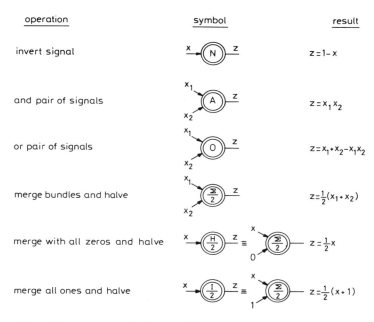

operation	symbol	result
invert signal	x —(N)— z	$z = 1 - x$
and pair of signals	x_1, x_2 —(A)— z	$z = x_1 x_2$
or pair of signals	x_1, x_2 —(O)— z	$z = x_1 + x_2 - x_1 x_2$
merge bundles and halve	x_1, x_2 —(Σ/2)— z	$z = \frac{1}{2}(x_1 + x_2)$
merge with all zeros and halve	x —(H/2)— z ≡ x, 0 —(Σ/2)— $z = \frac{1}{2}x$	$z = \frac{1}{2}x$
merge all ones and halve	x —(I/2)— z ≡ x, 1 —(Σ/2)— $z = \frac{1}{2}(x+1)$	$z = \frac{1}{2}(x+1)$

all operations are normalised to result in a standard bundle of H wires

$x = h/H$, $h =$ no. of '1's'

Fig. 1.8. *Bundle operations*

Ergodic processing simply combines bundle processing with the ideas of time stochastic processing, i.e. it uses the wires in a bundle as carriers of pulses. If the time average of the pulses on any wire is made equal to the average number of wires energised at any given time, we shall call such a bundle an *ergodic bundle*, or more simply an *ergodic*. Ergodics, by their very definition, have the remarkable property of carrying with them their own 'confidence indicator,' i.e. if time average (for any wire) and space average (for the bundle of wires) do not coincide, something must be fundamentally wrong. The interesting property of ergodics is that all the bundle processing operations are still valid; the ergodic property is processed automatically. This really means that the checking feature described above is nearly free of charge.

In order to discuss d.c. (analogue) signals, stochastic sequences, bundles and ergodics in their mutual relationship, it is useful to introduce the symbolism shown in Fig. 1.13. It should be noted that thin lines represent wires or electronic devices, while thick lines represent bundles and sets of electronic devices (attached to each wire of a bundle). To continue the usage introduced for bundles, there is an exception to this rule: logic symbols and the symbol for a difference amplifier are simply drawn double to indicate a bundle operation.

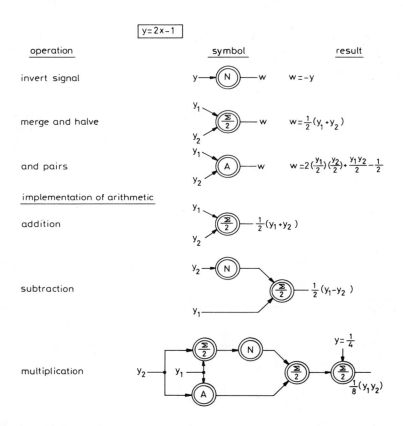

Fig. 1.9. *Remapped bundles*

Special mention should be made of the ergodic strobe (see Fig. 1.13). Here we simply have a sample-and-hold circuit on each wire. Strobing is done once, at an arbitrary time. The bundle synchronous scanner can be thought of as a disk carrying a contact (at the circumference) for each wire of the incoming bundle. A similar disk is connected to the outgoing bundle and rotates uniformly with respect to the first one. It is easily seen that this remaps any given input wire into all possible output wires in some sequence. The remarkable thing is that this remapping actually gives, on each output wire, a pulse sequence with a time average equal to the average over all wires at the input. This means that the synchronous scanner transforms

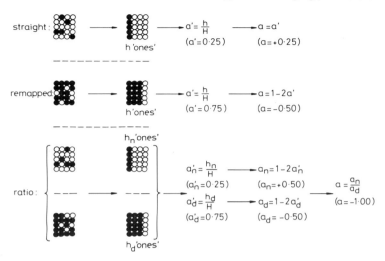

Fig. 1.10. *Straight, remapped and ratio bundle representations (all failsoft)*
Each bundle has H wires
a' = machine representation
a = actual value

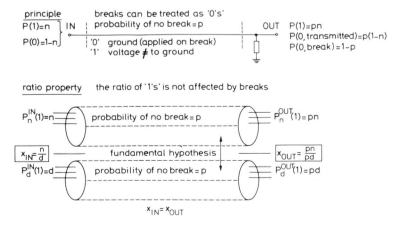

Fig. 1.11. *Failsafe bundles*

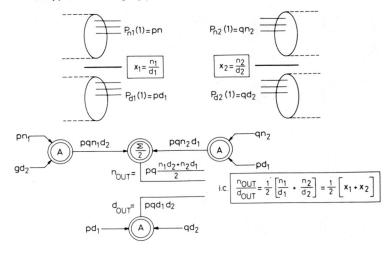

Fig. 1.12.   *Failsafe addition*

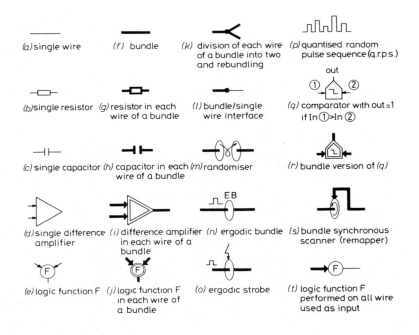

Fig. 1.13.   *Symbols for various processing schemes*

a bundle into an ergodic. Obviously, practical systems will replace rotating disks by circularly shifting registers.

Using the symbolism of Fig. 1.13 we can now draw an ergodic checking circuit in the form given in Fig. 1.14.

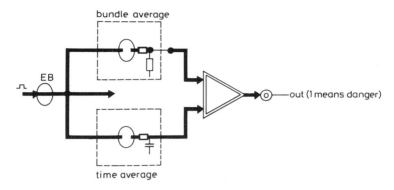

Fig. 1.14. *Ergodic checking circuit*

1.3 Burst processing

If it is desired to overcome the relatively slow gain of precision with the number of pulses in a completely random system like time stochastic processing, one can forego the minimum complication of stochastics and go back halfway toward a deterministic system. This leads to a new form of processing called burst processing, which strikes an interesting medium between determinism and a statistical interpretation of the pulse streams.

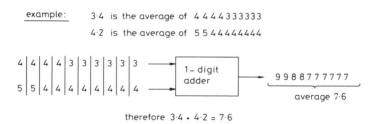

Fig. 1.15. *Fundamental idea of burst processing*

The idea of burst processing (see Fig. 1.15) is to *perform very low precision arithmetic* (typically on single decimal digits), *and to use appropriate averaging procedures to obtain higher accuracy.* Fig. 1.15 shows for instance how one can add 3·4 and 4·2 by decomposing 3·4 into a sequence of four 4's and six 3's, while 4·2 is decomposed into a sequence of two 5's and eight 4's. A one (decimal) digit adder then gives, as successive sums, two 9's, two 8's, and six 7's. The average of the latter is clearly 7·6, i.e. the sum of 3·4 and 4·2. The fundamental problem is obviously to produce automatically, in some circuit, the required integer sequences; we shall show below how this can be done.

Fig. 1.16 indicates some of the nomenclature and some of the conceptual extensions of burst processing. First, it is clear that any averaging procedure gives some error tolerance. This tolerance can be augmented if we decide on a representation of each integer (from 0 to 10, 10 included for reasons which will become clear below) which is unbiased, i.e. if we exclude such systems as classical weighted binary sequences (least significant digit first), as used in serial arithmetic units. In order to avoid difficulties in multiplication we map $0 \cdot n000 \ldots$ onto a 'burst' of n pulses in a 'block' of ten slots: we obtain an accuracy of $\leqslant 0 \cdot 1$. By averaging over ten blocks, a 'superblock', we obtain an accuracy $\leqslant 0 \cdot 01$. By averaging over ten superblocks, a 'hyperblock', we obtain an accuracy $\leqslant 0 \cdot 001$. For simplicity's sake we shall limit the discussion to superblock averages. The representation of negative numbers can be obtained by either signed bursts or a complementing system.

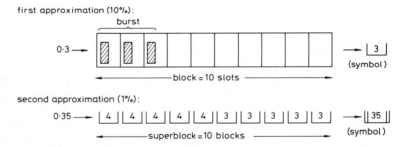

Fig. 1.16.   *Principle of burst representation*
The average over A:
Block = 10 slots: is exact to $\leqslant 0 \cdot 1$
Superblock = 10 blocks: is exact to $\leqslant 0 \cdot 01$
Hyperblock = 10 superblocks: is exact to $\leqslant 0 \cdot 001$

The design of burst processing systems may be accomplished using a few standard elements. One of these is the 'block sum register' (b.s.r.), which is simply a shift register of length 10, each bit driving a current source of strength V/R, where all R's are the same, but where V may vary from b.s.r. to b.s.r. The input clock in Fig. 1.17 should be construed to accept irregular gating pulses, as long as these do not occur at more than the clock frequency. A clear input allows initialising.

It is important to note that as long as we use the (summing) output of a b.s.r., i.e. as long as we essentially agree to 10-state logic we have a constant sum if the bursts remain equal (see Fig. 1.18). If we then design our system to use only the output of b.s.r.s in all arithmetic operations, it is no longer necessary to synchronise information coming in on separate channels, e.g. addend and augend. Thus burst processing transmits numerical data in a p.c.m. mode, is noise tolerant and does not necessitate synchronisation. Of course, the efficiency of the encoding in the 10-slot design (which could represent roughly 1000 numbers in weighted binary) is somewhat lower, since a block can only represent 10 numbers. In many applications this

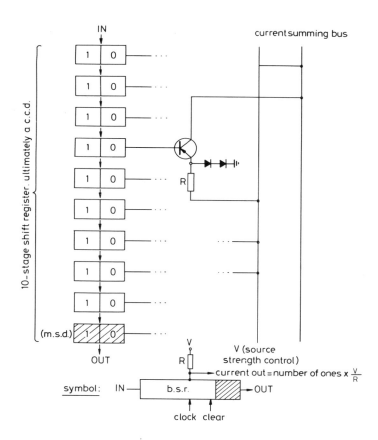

Fig. 1.17. *Block sum register*
The b.s.r. functions as an integrator over one block. Its indication is independant of the position of the burst

inefficiency is completely overshadowed by noise tolerance and nonsynchronisation. Note that it is, of course, entirely possible to use blocks of five or four slots, giving higher efficiency (4:16 ratio for four slots).

By using a stairstep encoder and a b.s.r. as a decoder as shown in Fig. 1.19 it is trivial to transmit audio and video signals by bursts. The interesting point is that the b.s.r. acts like an integrator which loses its memory after ten clock pulses. Furthermore, it is clear from the fact that burst processing is a form of p.c.m., multiplexing

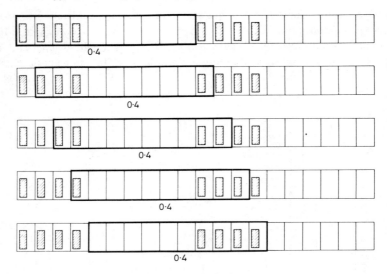

Fig. 1.18.   *Constant block sum property of periodic bursts*

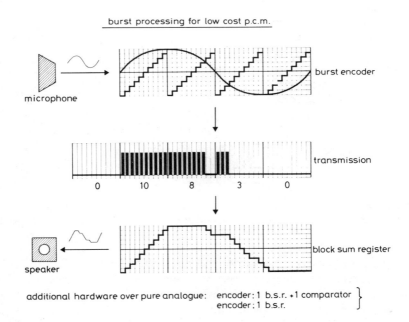

Fig. 1.19.   *Encoding and decoding of bursts*

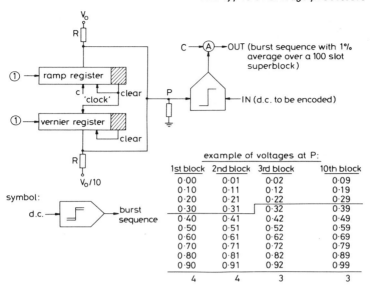

Fig. 1.20. *Vernier burst encoder*

example of voltages at P:

1st block	2nd block	3rd block	10th block
0·00	0·01	0·02	0·09
0·10	0·11	0·12	0·19
0·20	0·21	0·22	0·29
0·30	0·31	0·32	0·39
0·40	0·41	0·42	0·49
0·50	0·51	0·52	0·59
0·60	0·61	0·62	0·69
0·70	0·71	0·72	0·79
0·80	0·81	0·82	0·89
0·90	0·91	0·92	0·99
4	4	3	3

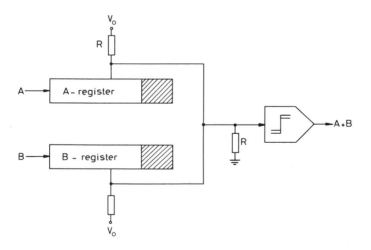

Fig. 1.21. *Block sum burst adder (subtractor)*
    — Scaling (overflow) can be obtained by modifying R
    — The superblock average of the output is equal to the sum of the superblock averages of A and B
    — Because a burst sequence is quasi periodic, the current output of the block sum register is a constant! Actually, we integrate without an RC circuit . . .
    — By taking current differences, we obtain a subtractor

is relatively easy. As a matter of fact, it is possible to attach to each block appropriate flags (a negative-pulse burst for instance) by which blocks can be sorted, redirected and reassembled. Again the absence of synchronisation problems makes the design much easier.

As mentioned above it is of paramount importance to produce the sequence of bursts which represent a number of higher precision than 0·1 by an automatic device. A simple circuit is a combination of two b.s.r.s successively filled with 'ones,' in which one b.s.r., the 'vernier register', runs at 1/10 the speed of the main b.s.r., i.e., the 'ramp register.' Each time the ramp register is cleared (when the 'ones' attain the most significant digit), the vernier register advances by one shift. As shown symbolically in Fig. 1.20, the vernier register adds to the output current of the ramp register one-tenth of one step of the former. Normalising the combined output current (or the voltage at P) to 1, the table in Fig. 1.20 shows how the slow shift upward of the steps produces first the longer bursts, then the shorter ones and precisely the requisite number. All that is necessary is to compare the voltage at P with the (d.c.) voltage to be encoded (also normalised to 1) and to switch on the clock pulses by an AND gate as long as the voltage to be encoded (0·32 in Fig. 1.20) is higher than the comparison voltage.

It is to be noted that in case of many voltages to be encoded in one location, one vernier encoder is sufficient; since the result of the encoding, i.e. the burst sequence, is always assessed by an integrating b.s.r., no nefarious correlation effects can possibly occur.

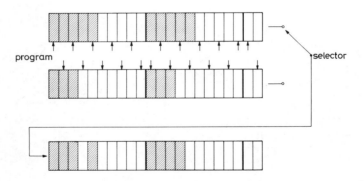

Fig. 1.22.   *Selector adder*
— The selector is programmed to switch phase at the end of each block
— The output is uncompacted and must be used to shift ones into an assembly register

Fig. 1.21 shows the extremely simple layout of an adder. It consists of two b.s.r.s working into a common bus, the result being re-encoded by a vernier encoder (closer analysis shows that actually a fixed 10-step encoder is sufficient.) Note that this figure represents a 10-state logic circuit. It is generally felt that the simplifications that can be obtained by going to multistate logic are now within the reach of semiconductor technology. Nevertheless, it should be emphasised that one can operate without b.s.r.s in the processing sections. Fig. 1.22 shows a so called selector adder for compacted bursts (Note the analogy with the stochastic adder in Fig. 1.4). In order to eliminate bias for odd numbers of pulses, the selector switches phase at the end of each block. The output is uncompacted, and in order to obtain a compact output we shift 'ones' into an assembly register whenever a 'one' occurs in the bottom sequence. A more detailed treatment of other burst processing arithmetic operations and applications will be presented in Chapter 5.

Chapter 2

Fundamental theory of stochastic processing

The basic idea of stochastic processing was originated independently and almost simultaneously in the USA and the UK.[3,4] A review paper subsequently considered some of the fundamental ideas and possible applications.[13] The application of stochastic processes to computer design was first suggested in a classic paper,[1] and the technique of obtaining multiplication by pulse coincidence had been demonstrated.[2] Since this early work stochastic processing has become a subject of investigation in various research centres throughout the world.[14] Research groups now exist in the USA, UK, France, Germany, Japan, Spain and several other countries.

As we have seen in Chapter 1, the variables in time stochastic processing are represented by sequences of digital pulses, the information being carried by the probability of occurence of an ON logic level. Each logic level is generated from a random variable, and the statistically independent results form a Bernoulli sequence whose average pulse rate is determined by the variable to be represented. It should be noted that there is no quantisation error with this form of probability representation, but the variable cannot be measured exactly from a sequence of finite length because of the variation inherent in the generation of a Bernoulli sequence.

In some applications of stochastic processing, e.g. the stochastic learning automata to be discussed subsequently, no input interface problems arise because we are dealing with direct computations involving random sequences. In general, for applications involving deterministic variables, it is necessary to map the input variables to within the $(0,1)$ range of the probabilistic representation. Several such mappings have been proposed,[15] and in this Chapter we consider in particular a bipolar mapping which leads to simple hardware realisations for the various stochastic operators. Because of the relationship between Boolean algebra and discrete probability theory we show that provided conditions of statistical independence are satisfied the normal arithmetic operations of inversion, multiplication, addition and integration may be performed by simple logic circuits. Since random variables can only be estimated with a certain expectation of error, the study of errors is particularly important. For each stochastic arithmetic operation we derive expressions which permit the prediction of error.

The validity of the stochastic circuits relies heavily on the assumed property of statistical independence between operating variables. Hence, of vital importance is the design of suitable *random number generators* for the provision of independent, uniformly distributed, random numbers. We consider the various approaches to the problem of random number generation, and describe a particularly convenient technique using negatively correlated pseudorandom binary sequences.

The output of a time stochastic processor will generally be in the form of a non-stationary Bernoulli sequence. Such a sequence can be considered in probabilistic terms as a deterministic signal with superimposed noise. The output interface of the system must be able to reject the noise component and provide a measure of the mean value of the observed sequences generating probability. In addition, the interface must be amenable to reasonably simple synthesis with digital logic circuits. The Chapter concludes with a discussion of optimum criteria which may be used for the evaluation of output interface systems and gives examples of possible circuit realisations.

2.1 Input mappings

The actual mapping used substantially influences the simplicity of the hardware required for a specific computation. We will only consider linear mappings, although it should be noted that nonlinear mappings exist which permit computations with numbers in an infinite range with logarithmic error characteristics.[15, 16] The simplest form of linear mapping is the *unipolar transformation* in which some physical quantity E in the range, $0 \leqslant E \leqslant V$, represents a binary variable, A, with probability p such that

$$p = p(A = 1) = \frac{E}{V}$$

Thus the maximum value of the range $E = V$, is represented by a logic level always ON, $p(\text{ON}) = 1$, and zero value is represented by a logic level always OFF, $p(\text{ON}) = 0$. Intermediate values of E will be represented by a randomly fluctuating logic level with a certain probability of being ON at any instant.

Let the value of the binary variable A be noted at every clock interval with A_i being the value at the ith clock interval. The expected value of A_i is:

$$\text{Exp}[A_i] = p$$

An estimate of the generating probability, $\hat{p}$, taken over N clock intervals is given by

$$\hat{p} = \frac{1}{N} \sum_{i=1}^{N} A_i$$

The expected value of the estimate, $\text{Exp}[\hat{p}]$, is

$$\text{Exp}[\hat{p}] = p$$

The *variance* of the estimate of $\hat{p}$ is by definition the expected value of the square of the difference between the estimated value of the mean of A_i and the expected value of A_i

Thus

$$\mathrm{var}(\hat{p}) = \mathrm{Exp}[(\hat{p} - p)^2]$$

$$= \frac{1}{N^2}\mathrm{Exp}\left[\sum_{i=1}^{N}(A_i - p)^2 + \sum_{i=1}^{N}\sum_{j=i+1}^{N}(A_i - p)(A_j - p)\right]$$

The last term of this equation is zero since a Bernoulli sequence implies that

$$\mathrm{Exp}[(A_i - p)(A_j - p] = 0$$

Hence

$$\mathrm{var}(\hat{p}) = \frac{1}{N^2}\mathrm{Exp}\left[\sum_{i=1}^{N}(A_i - p)^2\right] = \frac{\mathrm{Exp}(A_i^2) - p^2}{N}$$

Since $A_i^2 = A_i$

$$\mathrm{Exp}(A_i^2) = \mathrm{Exp}(A_i) = p$$

Finally we have

$$\mathrm{var}(\hat{p}) = \frac{p - p^2}{N} = \frac{p(1 - p)}{N}$$

Hence the expected value of the estimate is independent of the number of clock intervals, but the accuracy of the estimate must be determined by the variance which is a function of N. The standard deviation, σ, of the estimate of p is given by:

$$\sigma(\hat{p}) = \left\{\frac{p(1 - p)}{N}\right\}^{1/2} \tag{2.1}$$

Thus the expected error is zero when p is equal to zero or one, corresponding to deterministic sequences. Maximum error occurs when $p = \frac{1}{2}$. *The error is inversely proportional to the square root of the sample size or number of clock intervals N.* In other words the longer the sample, the less the error, but the longer is the time needed for making measurements. An illustration of the usual classic compromise between computational speed and accuracy. It should be noted that for any measured value of probability p there is a 0·68 chance of the value lying within the range $p \pm \sigma$. Hence full knowledge of the output can only come from a complete description of the variables probability distribution function. However, as a convenient way of summarising the information and to gain a simple measure of accuracy, the statistics of the mean and standard deviation are quoted.

In the two-line bipolar mapping the unipolar representation is extended to bipolar quantities (both positive and negative) by using two sequences of logic levels on separate lines, one representing positive quantities and the other negative.[15] The line whose probability is weighted positively is called the UP line and the negatively weighted line is called DOWN. For a given variable E within the range $-V \leqslant E \leqslant +V$ we let

$$\frac{E}{V} = p(\text{UP line ON}) - p(\text{DOWN line ON})$$

Thus maximum positive quantity is represented by the UP line always ON and the DOWN line always OFF. Maximum negative quantity occurs when the UP line is always OFF and the DOWN line always ON. For intermediate quantities stochastic sequences will occur on one or both lines. In particular, zero will be represented by equal probabilities of an ON logic level for both lines or by both lines being OFF.

The mapping of most interest to the present work is the single-line bipolar mapping in which a variable E in the range $-V \leqslant E \leqslant +V$ is represented by the mapping:

$$p(\text{ON}) = p(A = 1) = \frac{1}{2} + \frac{1}{2}\frac{E}{V} \tag{2.2}$$

Clearly for maximum positive quantity $E = V, p(\text{ON}) = 1$, and for maximum negative quantity $E = -V, p(\text{ON}) = 0$. Zero is represented with an equal probability of being ON or OFF.[15]

We now consider the digital logic required to implement the basic mathematical operations using the single-line bipolar mapping. Although the operation of all the circuits we consider is synchronous being governed by a master clock, it should be noted that asynchronous time stochastic techniques have also been studied.[12]

2.2 Basic stochastic computing

2.2.1 Inversion (complementation)
A simple logic invertor gate, Fig. 2.1*a*, gives multiplication by -1. An *invertor* complements an input sequence so that the binary input variable A, is related to the output A_0, by:

$$A_0 = 1 - A_1$$

Using expected values gives

$$\text{Exp}[A_0] = \text{Exp}[1 - A_1]$$

Hence

$$p_0 = 1 - p_1$$

By substituting in eq. 2.2 we obtain

$$E_0 = -E_1 \tag{2.3}$$

The measure of standard deviation used previously in eq. 2.1 related the error in estimation of the machine variable p to the full scale range of that variable. It is more meaningful to express the error in terms of the normalised particular value of the input variable E/V. If we introduce ϵ as the new measurement of error we obtain:

$$\epsilon = \frac{\sigma(E/V)}{E/V} \tag{2.4}$$

where $\sigma(E/V)$ represents the standard deviation of the normalised input variable.

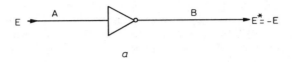

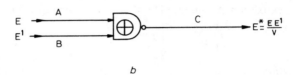

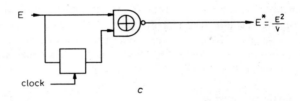

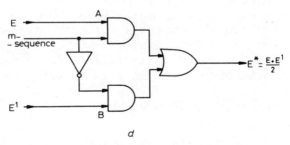

Fig. 2.1.   *Stochastic computing elements*
   a Invertor
   b Multiplier
   c Squarer
   d Summer

From the single-line bipolar transformation given by eq. 2.2 we have

$$\frac{E}{V} = 2p - 1$$

which implies that

$$\frac{\hat{E}}{V} = 2\hat{p} - 1$$

Thus the standard deviation of $\hat{E}/V$ is twice the standard deviation of $\hat{p}$. Directly from eq. 2.4. we have

$$\sigma\left(\frac{\hat{E}}{V}\right) = 2\left\{\frac{p(1-p)}{N}\right\}^{1/2} \tag{2.5}$$

Note that the standard deviation is a maximum when $p = 0\cdot5$ or equivalently $E = 0$. Considering now the invertor circuit the output is a Bernoulli sequence with the same standard deviation as the input. This is verified from eq. 2.5 which shows that $\sigma(\hat{E}/V)$ is independent of the sign of E.

2.2.2 Multiplication

To multiply two quantities, an inverted exclusive-OR gate may be used as shown in Fig. 2.1b. The expected value of the output Exp $[A_0]$ may be related to the binary inputs A_1 and A_2 by:

$$\text{Exp}\,[A_0] = \text{Exp}\,[A_1 A_2 V \bar{A}_1 \bar{A}_2]$$

where $\bar{A}_1 = 1 - A_1, \bar{A}_2 = 1 - A_2$ and the symbol V represents logical OR.

   In terms of probabilities we may write

$$p_0 = p_1 p_2 + (1 - p_1)(1 - p_2) + 2\,\text{covariance}\,(A_1 A_2)$$

Assuming that the covariance term is zero we may use the mapping of eq. 2.2 to obtain

$$E_0 = \frac{E_1 E_2}{V^2} \tag{2.6}$$

Thus we obtain *normalised multiplication* of the input variables. It should be observed that satisfactory multiplication is obtained only with the implicit assumption that the covariance term is zero. The validity of this assumption is dependent on the statistical characteristics of the random number sources used to generate the input sequences.

   Assuming zero covariance, it may be shown that the variance of the output probability is given by[17]

$$\sigma^2(p_0) = (p_1 p_2 + q_1 q_2)(1 - p_1 p_2 - q_1 q_2)$$

where $q_1 = 1 - p_1$ and $q_2 = 1 - p_2$.

   Hence, provided that there is no correlation between inputs, the variance of the output probability is entirely governed by the value of the output. This indicates that the error in any calculation is independent of the number of multiples used, and, with no accumulation of error, the accuracy is just a function of the output sequence probability. However, if two input probabilities are multiplied, the product is always less than the individual input probabilities. Thus, if stochastic sequences pass through a cascaded arrangement of multipliers, the probability associated with the resultant sequence is successively attenuated.

As we have stressed, all stochastic processing elements depend on the fundamental assumption that input sequences can be considered statistically independent. It follows directly that to square a number it is not sufficient merely to short circuit the inputs of a multiplier. Fortunately, in a Bernoulli sequence each event is independent from any other event. This means that if a stochastic sequence is put alongside a one-bit delayed replica of itself, the two sequences will appear to be independent with no cross-correlation at time $t = 0$. The circuit for a squarer is therefore a multiplier with some form of flip-flop delay on one input as shown in Fig. 2.1c.

2.2.3 Addition

To obtain stochastic sequence summation it is tempting to simply pass the sequences through a logic OR gate. As explained in Chapter 1, the situation is complicated by both the possible presence of coincident pulses and the problem of scaling. If two input sequences of probabilities p_1 and p_2 are passed through an OR gate the output probability is given by:

$$p_0 = p_1 + p_2 - p_1 p_2$$

Thus, true summation is only achieved when $p_1 p_2 = 0$, or when the two inputs are mutually exclusive. In addition, the input probabilites must be such that $p_0 \leqslant 1$. The above problems may be solved by randomly connecting either of the two binary inputs A_1 and A_2 using the circuit of Fig. 2.1d. The output of this circuit will again be a Bernoulli sequence of probability p_0 such that

$$p_0 = \tfrac{1}{2}(p_1 + p_2)$$

Assuming zero covariance the variance of the output probability will be

$$\sigma^2(p_0) = \sigma^2(\tfrac{1}{2}[p_1 + p_2])$$

Again we observe that provided the covariance terms are zero, the variance of the output sequence is purely a function of the output probability p_0. Using the bipolar transformation of eq. 2.2 we now have

$$E_0 = \tfrac{1}{2}(E_1 + E_2) \tag{2.7}$$

It is of interest that in this case correlation between the two inputs E_1 and E_2 would not invalidate the summation operation. Provided that the switching probability of 0·5 is statistically independent from the inputs, the circuit will perform addition, even with deterministic inputs.

At this stage a general conclusion may be made. The results for stochastic processing involving inversion, multiplication and addition demonstrate that stochastic processors synthesised using combinational logic do not introduce any intrinsic calculation errors.

2.2.4 *Integration*

The discussion so far has been exclusively concerned with arithmetic operations which may be simply realised using combinational logic circuitry. Integration, however, with its adding and storing operations, requires that the signal is memorised, and for this purpose flip-flops are connected in cascade to form counters. In a stochastic processor, a probability measure is integrated over a time period T, by counting the number of pulses which occur in the sequence over the time period. To be able to integrate a variable E in the range $-V \leqslant E \leqslant V$ we must accomodate negative quantities. Physically, this means that the counter must be able to count down as well as up. The counter may be arranged to increase by unity if the increment line is ON and the decrement line OFF. The counter counts down if the opposite condition applies, and if the inputs are both ON or OFF simultaneously the counter remains in the same state.

Suppose the counter has $(n + 1)$ states labelled $\pi_0, \pi_1, \pi_2, \ldots \pi_n$. The output of the integrator is π_i when it is in the ith state. If π_i lies within the range $(0, 1)$ then

$$\pi_i = \frac{i}{N}$$

Let w be the probability that the integrator will count up at the next clock pulse and e be the probability of count down.

The expected change at the nth clock pulse is given by:

$$\delta\pi(n) = \left(\frac{w(n) - e(n)}{N} \right)$$

Where $1/N$ represents a one step change. Over a time interval of m clock periods, the expected change would be

$$\pi(m) - \pi(0) = \sum_{n=0}^{m-1} \delta\pi(n) = \sum_{n=0}^{m-1} \left(\frac{w(n) - e(n)}{N} \right)$$

The right-hand side of this equation is a simple zero-order numerical integration formula for $w(n) - e(n)$, and the equation can be rewritten as

$$\pi(m) = \pi(0) + \frac{1}{N} \int_0^{m-1} \{w(n) - e(n)\} \, dn$$

Thus the up/down counter may be regarded as an integrator whose real-time gain depends on the number of counter states N, and the frequency of the master clock. For a clock period T the above equation may be rewritten as

$$\pi(\tau) = \pi(0) + \frac{1}{NT} \int_0^\tau \{w(t) - e(t)\} \, dt$$

where τ will be some integer multiple of T.

To integrate a variable in the bipolar representation the input is connected directly to the increment line, and in inverted form to the decrement line, as shown in Fig. 2.2. The probabilities w and e are then given by the equations

$$w = p_1$$

$$e = 1 - p_1$$

and hence using the bipolar mapping eq. 2.2 we have

$$E_0(\tau) = E_0(0) + \frac{2}{NT} \int_0^\tau E_1(t)\,dt \tag{2.8}$$

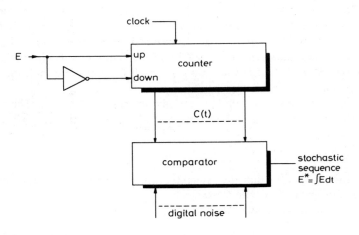

Fig. 2.2.   *Single input integrator*

Note that the effective gain of the integrator is doubled. It is also possible to design an integrator whose expected output is the integral of the sum of the inputs. The basic circuit is shown in Fig. 2.3. We now have

$$w = p_1 p_2$$

$$e = (1 - p_1)(1 - p_2)$$

and hence using eq. (2.8) gives

$$E_0(\tau) = E_0(0) + \frac{1}{NT} \int_0^\tau [E_1(t) + E_2(t)]\,dt \tag{2.9}$$

The problem of drift in integrators provides a significant potential source of error in stochastic processing. Consider the simple single-input integrator of Fig. 2.2. Although because of the digital representation stochastic integrators do not suffer from conventional analogue circuit drift, errors are caused by the process of integration which accumulates all input logic levels over the experimental time. Thus, for example, a long sequence of logic 1's can orientate the counter into a biased state such that the bias can only be neutralised by a similar sequence of logic 0's. After a given period of integration there is a high probability that the counter will move into a position above or below the true mean and remain at this position for the remainder of the experiment. If the integrator states are sampled at regular time intervals, as part of a read-out operation, the computed average will tend to represent a biased estimate of the mean. Apart from the use of negative correlation, to be discussed subsequently, it is possible to reduce drift by a modification of the integrator structure.[18]

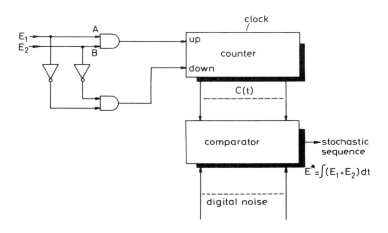

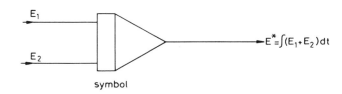

symbol

Fig. 2.3.   *Summing integrator*

We consider the state of an integrator to be a random variable π whose mean and variance depends on the probability of the input sequence. Suppose π_0 represents the initial state of the integrator and π_N be the state after N clock pulses. The mean and variance of π are given by:

$$\text{Exp}\,[\pi] \;=\; N(p-q) \tag{2.10}$$

$$\{\sigma(\pi)\}^2 \;=\; N\{(p+q)-(p-q)^2\} \tag{2.11}$$

where p and q represent the probabilities of count-up and count-down, respectively. If $p \neq q$, the integrator has a steadily changing mean and the variance will depend on both the sum and difference of the probabilities p and q. If $p = q$ the mean value becomes the integrators initial state π_0, and the variance is solely a function of the sum of the probabilities.

Irrespective of the actual value of p and q the error after N clock intervals will be of the order of $\sqrt{N}$ states. Thus, the longer the solution time the greater the probability of the integrator making extensive excursions from the mean. Eq. 2.11 indicates that accuracy is adversely affected by the integrators capacity of counting down. Eq. 2.10 implies that in order not to change the estimated mean value of the sequence, any reduction in q to reduce the variance must be accompanied by an equal reduction in p. Now since $p + q = 1$ we have

$$p - q = p(1 - q) - q(1 - p) = p^2 - q^2 \tag{2.12}$$

This equation indicates that if the count-up and count-down probabilities are p^2 and q^2, respectively, the mean estimate will be unaffected. The new variance will be given by

$$\{\sigma_1(\pi)\}^2 = N\{(p^2 + q^2) - (p - q)^2\} = \{\sigma(\pi)\}^2 - 2Npq$$

Thus, although the mean is unaffected, the variance is reduced by the factor $2Npq$. This procedure leads to a general technique which may be repeated until one of the probabilities is arbitrarily close to zero. Thus

$$p - q = p^2 - q^2 = P^2(1 - q^2) - q^2(1 - p^2) \tag{2.13}$$

The variance in this case is

$$\{\sigma_2(\pi)\}^2 = \{\sigma(\pi)\}^2 - 2pq(1 + pq)N$$

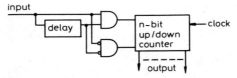

first circuit modification

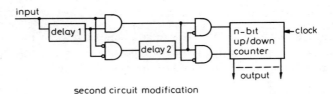

second circuit modification

Fig. 2.4.   *Modified integrator structures*

As a simple numerical example suppose $p = q = \frac{1}{2}$. This gives:

$$\sigma(\pi) = (N)^{1/2}, \ \sigma_1(\pi) = (0 \cdot 5N)$$

and

$$\sigma_2(\pi) = (0 \cdot 375N)$$

The digital circuits used to implement eqs. 2.12 and 2.13 are shown in Figs. 2.4*a* and 2.4*b*. Delay circuits are included to ensure that all input sequences are independent. If the delay 1 equals the delay 2 the input sequences will be cross-correlated.

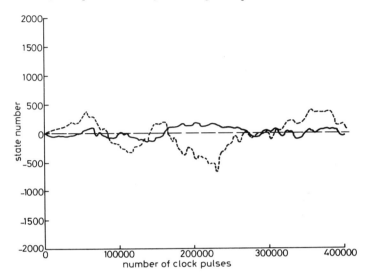

Fig. 2.5. *Experimental results for modified integrator*
- - - - - - - basic integrator
——————— modified integrator
input probability = 0·5

It should be noted that Fig. 2.4*a* is essentially the same circuit as the 2 input summing integrator of Fig. 2.3 modified by the inclusion of a delay. Typical experimental results are given in Fig. 2.5 which shows results for a standard single-input integrator of Fig. 2.2 compared with the modified circuit of Fig. 2.4*a*. As predicted, the second circuit provides a decrease in variance. In fact irrespective of the input probability the circuit of Fig. 2.4*a* will provide one half of the variance of Fig. 2.2.

2.2.5 Other arithmetic operations

It may be shown[12] that division can be accomplished with little circuitry. However, the result obtained is unfortunately no longer random, i.e. bunching occurs because of the form of algorithm used. In practical designs it is usual to adopt some form of two-line mapping. Essentially, division is a somewhat difficult problem because of the finite range of probability. Thus division by a small enough quantity can lead to a result outside the range of representation. It is possible to design circuits for both division, square-root extraction and other function generation using dynamic error reducing procedures based on steepest descent techniques.[15]

2.3 Random number generation

All the methods for stochastic sequence generation follow the same basic principle. The deterministic input D is mapped into the range $(0, 1)$, and then compared with a random number N which takes on any value within the limits 0 and 1. The output signal of $D > N$ is a Bernoulli sequence, where the probability of an ON logic level is directly dependent on the magnitude of the input signal D. For there to be a linear relationship between the input D and the resulting probability p of the corresponding stochastic sequence the probability density function $f(x)$ of the random number must be a constant since

$$p = \int_0^D f(x)\, dx$$

The random numbers must therefore be uniformly distributed across the specified range of the inputs.

As we have seen the validity of using simple logic gates for multiplication and addition relies heavily on the assumed statistical independence between input variables. Hence, one of the major problems in the design of stochastic processors is the synthesis of hardware systems for the generation of sequences of independent, uniformly distributed, random numbers. Early forms of random number generators for stochastic processors used physical noise sources such as microplasma diodes. Fig. 2.6 shows the basic method of comparing a noise waveform with an external input signal using a standard analogue comparator.[12] The resultant pulse train is sampled at regularly spaced intervals to avoid any correlation. This form of noise generator has been extensively investigated both in Germany and in France.[19-21] The main advantage of such a system is that a combination of natural physical noise sources is independent over an infinite period of time (assuming that the external environment does not induce correlations).

The main disadvantages of analogue noise sources are as follows:

(*a*) The noise signals cannot be duplicated. This is a serious restriction, particularly in computations involving system simulation. In addition, if a system breakdown occurs, fault finding would become extremely difficult, since repetition is one of the most important techniques of troubleshooting, especially in the random environment of a stochastic system.

(*b*) The physical noise sources can be very sensitive to environmental changes, and to overcome this, feedback circuitry must be incorporated. Essentially, the problem is to maintain the mean level of the waveform constant and to ensure that the bandwidth is high enough so that regular sampling provides statistically independent results.

(*c*) Most analogue circuits rely on some form of sampling to guarantee independence. This tends to limit the speed of number generation.

It is known that Gaussian distributed diode noise sources give nonlinear input interface characteristics, especially when the noise standard deviation is kept small to achieve minimum output variance. One possible technique for reducing non-linearity is to convert each input sample twice, the second time with an offset of half a quantisation step, and finally calculate the arithmetic mean of the two outputs.[22]

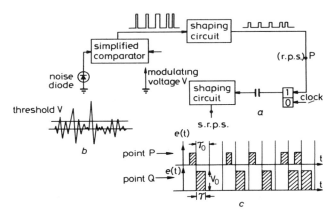

Fig. 2.6. *Analogue noise generation*

An alternative approach to random number generation is to consider the use of the well known pseudorandom binary sequence generators.[23-25, 82] The objective is to produce a sequence of determined logic levels, which will behave over a finite time interval as though generated by a truely random process. The sequences are formed in shift registers with appropriate feedback.

Consider a sequence of logic levels of length N circulating within an n-bit shift register as shown in Fig. 2.7. To appear random the sequence must pass the following tests which characterise a Bernoulli sequence of probability 0.5:

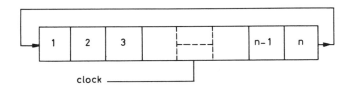

Fig. 2.7. *N-bit circulating shift register*

(*a*) For a sequence of N logic levels, the number of ON levels should equal the number of OFF levels. If N is an odd number, the difference between the ON and OFF levels should not exceed one.

(*b*) The number of runs of consecutive logic levels, either ON or OFF, should be directly related to the length of the run. Half the runs should be of length 1, a fourth of length 2, an eighth of length 3, and so on.

(*c*) For all delays within the sequence except the zero delay, the value of the auto-correlation function of the sequence must be kept small.

A logic sequence will pass tests (*a*) and (*b*) if when passed down a short *n*-bit shift register, it causes the register to go through all of its 2^n possible states in 2^n clock periods. Hence, if the number of stages, *N*, of the cycling shift register is made equal to 2^n, then a sequence can be devised and loaded into the shift register which will satisfy the first two conditions of randomness. Unfortunately, this approach of storing the entire sequence ceases to be possible with practical values of *N*. An alternative approach is to generate the complete cycle of $2^n - 1$ numbers (*m* − sequence) by performing simple arithmetic operations on the *n*-bit number stored in a shift register by means of exclusive-OR feedback circuits (Fig. 2.8). It is well known that such configurations may be described in terms of a characteristic equation given by

$$x^n \oplus x^{n-s} \oplus 1 = 0 \tag{2.14}$$

where *x* is a delay operator, *n* the length of the shift register and *s* is the feedback stage. For a maximum-length sequence of $(2^n - 1)$ states to be generated, the characteristic polynomial must be primitive. Standard tabulations exist for irreducible polynomials simplifying the design of maximum length shift register sequences.[26]

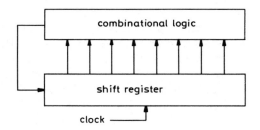

Fig. 2.8.   *N-bit shift register with feedback*

The use of stochastic sequences in forming random numbers must be treated with caution. It is tempting to assume that, with the logic levels in adjacent clock periods being independent random variables, *n*-bit random numbers will result from shifting an *m*-sequence along an *n*-bit shift register as shown in Fig. 2.9. For example if, owing to the first stage being a binary '1', the random number is greater than 0·5, the following number is certain to be above 0·25, when the binary '1' is shifted to the second stage. Hence consecutive numbers would not be independent. Alternatively, if the register is clocked *q* times between the numbers being taken, the independence can be assumed. The value of *q* must be greater than or equal to the number of bits of the register, and prime to the length *N* of the *m*-sequence.[27] Thus the sequence of ones and zeros making up the individual digits of the binary numbers are formed from taking every *q*th term of the *m*-sequence. The first digit has the terms $1, q + 1, 2q + 1$ etc and the nth digit has the terms $n, q + n, 2q + n$ etc., The n separate sequences are the replicas of a single m-sequence, but delayed approximately N/q clock pulses from each other.

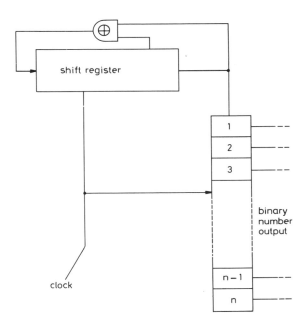

Fig. 2.9. *Random number generator with positive correlation*

Consider the example of M random-number sequences each having a word length of n binary digits. To avoid duplication of the sequences within a period t of processing time, the bit length of each sequence must be tf_c, where f_c is the system-clock frequency. Thus the length of the m sequence N is given by

$$N = Mn \cdot t \cdot f_c$$

The delays will be integer multiples of $(N/M)n$ which might be of the order of 10^9 clock periods. This simple example illustrates that the generation of delayed sequences is a far more complex problem than the generation of the original m-sequence given by eq. 2.14. The equation can be written in generalised form as:

$$x^d = x^{d-s} \oplus x^{d-n} \tag{2.15}$$

where d can take on any value between 0 and $2^n - 1$. Thus a higher order delay can be expressed as the sum of two lower order delays, and by repeated use of the equation any delay can be reduced to the modulo -2 sum of delays in the range 1 to n, which are directly available from the generating shift register. Each time eq. 2.15 is applied, the number of sequences to be added together is doubled, and if the process is drawn out in graphical form, it takes on the pyramidal structure shown in

Fig. 2.10. The number of each layer denotes the number of times eq. 2.15 has been used. For large delays the pyramid can become very complex but, owing to cancellation, the rows do not always double in size as eq. 2.10 is applied. In Fig. 2.10, for example, the second and third numbers in row 2 are both the result of subtracting $(n + s)$ from row 0. Since the output of an exclusive-OR gate is zero for equal inputs, only duplicated delays are erased from the pyramid. Fig. 2.11 shows the modification of Fig. 2.10 with cancelled delays owing to duplication. A large scale graph (Fig. 2.12) shows the development of regular cancellation patterns with the formation of triangular areas of erased sequences.[28]

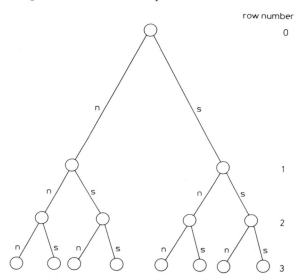

Fig. 2.10.   *Pyramid structure of delayed sequences*

When the row number is a power of 2 the sequences are mainly duplicates of each other and only the two extreme delays at either end of the row are left uncancelled. When the row numbers are integer multiples of 4 there is always reduction, and the same applies to a lesser extent with multiples of 2. This is not an altogether unexpected pattern. The elements in row $(i + 1)$ are formed by subtracting two numbers, n and s, from each element in the ith row. The $(i + 1)$th row is therefore derived by subtracting $(nx + sy)$ from the source delay at the point of the pyramid, row 0, where $x + y = i + 1$. The various delays constituting a row originate from all the possible combinations of x and y which add up to $(i + 1)$. Cancellation occurs when there is an even number of permutations of n and s taken $(i + 1)$ at a time.

We now have a simple method of calculating which of the n stages of the generating shift register must be exclusive-ORed to produce a high order delay. A search is first made for the lowest row which is numbered 2^s where s is an integer, and which satisfies the equation.

$$r - 2^s n \geqslant 0 \qquad \text{where } n > s$$

Having found the largest value of s which satisfies this equation, the two uncancelled delays are evaluated as:

$$d_1 = r - 2^s n \qquad d_2 = r - 2^s s$$

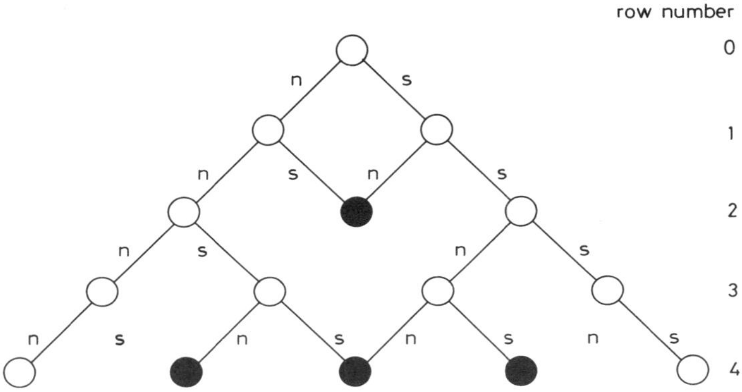

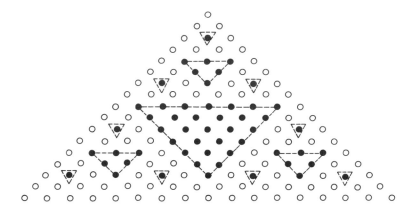

Fig. 2.11.   *Sequence cancellation of pyramid structure*

Fig. 2.12.   *Graph showing formation of cancellation patterns*

These two values, d_1 and d_2, are now treated as two separate source delays and two new values of s are computed. The procedure is repeated until all delays are less than the length of the generating register. Many of them will be duplicated and thus can be cancelled. Those delays which are not repeated or are repeated an odd number of times are retained, and form the solution. Such an algorithm is simple to program and can handle delays of several million.

Fig. 2.13 shows the schematic diagram of a random number generator designed according to the above theory. From a central 31-bit shift register the $2^{31}/_{32}$ delay is

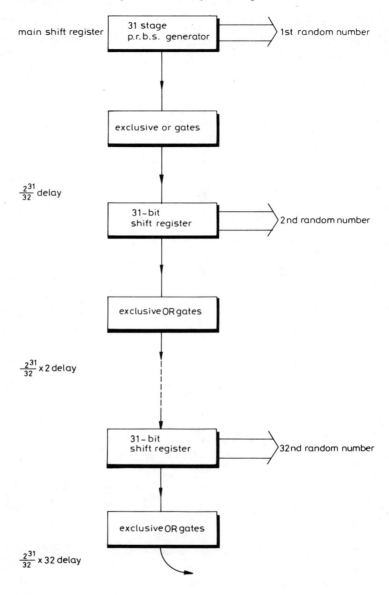

Fig. 2.13.   *Schematic diagram for random number generation*

obtained using cascaded exclusive-OR gates. This delay is connected into a second 31-bit shift register whose outputs are exclusive-ORed in exactly the same manner as the central shift register to provide the delay $(2^{31}/_{32}) \times 2$. As shown in Fig. 2.13, this arrangement may be duplicated to obtain 32 sequences. This basic generator has been subjected to a wide variety of successful statistical tests and theoretical analysis.[27]

An alternative approach to the generation of independent Bernoulli sequences is to provide a switched arrangement of equal length m-sequences scrambled by one common m-sequence.[29] Such an approach has been shown to permit the generation of 25 random sources with zero cross-correlation up to a time delay of 60 hours at a clock rate of 10 MHz. Theoretically, the number of sources may be extended to 4 million without affecting the statistical parameters.

2.4 Output interface design

The basic output interface or stochastic estimation problem is illustrated in Fig. 2.14. A deterministic output is required to be produced corresponding to the probability of occurence of an input pulse. The properties governing the efficiency of output interfaces are similar to those of conventional statistical estimation, and may be summarised as shown below.[30]

stochastic input
sequence

optimal
estimator

deterministic
output

Fig. 2.14.   *Basic estimation problem*

Steady-state characteristics:

(*a*) *Minimum bias error:* Ideally, the output interface should provide an unbiased estimate with the expected value of the output equal to the mean of the measured variable.

(*b*) *Minimum variance:* The variance of the output should be commensurate with the length of the sample taken.

Step-response characteristics:

(*c*) *Minimum response time* to minimum bias error. Obviously, the time taken to reach an unbiased estimate should be as short as possible.

(*d*) *Minimum response time* to minimum variance: The response time should equal the theoretical minimum time required to compute an estimate with the desired variance.

Nonstationary input characteristics: For nonstationary inputs, variants on all previous properties will exist, depending on the nature of the nonstationarity. One important characteristic is the following:

(e) *Ability to track nonstationary inputs:* Ideally, the output interface should be able to continually track a changing input value without the necessity of resetting the circuit after each reading.

The ideal output interface for a stochastic system is one such that the normalised variance decreases with time and all readings form unbiased estimates of the input probability. Thus the most significant bit of the binary-number estimate would stabilise first, followed by progressive stabilisation of further bits as the normalised variance decreases. An arithmetic operation such as this requires continual division and is beyond the bounds of a simple digital realisation.

If N consecutive clock intervals of a Bernoulli sequence are examined, and the number of ON logic levels counted, then the ratio of the count to the number of clock intervals gives an estimate of the sequences generating probability. When the sequence represents a fixed quantity, so that the corresponding Bernoulli sequence is stationary, the accuracy of the measurement can be increased by increasing the sample size N. The price paid for the increase in accuracy is the length of time required to make the measurements. With a time-varying input signal, the output interface must have the ability to track the signal continuously, or else the higher-frequency components of the waveform will be lost. To do this requires the calculation of a short time or *Moving Average*. [31]

To form a Moving Average of a stochastic sequence, the presence or absence of a pulse in all of N adjacent clock periods is recorded and stored, and the average pulse rate calculated. The next clock interval of the sequence is then interrogated and the average is recalculated over the N most recent clock periods. The information contained in the first interval is lost. If p_N is the estimate of the generating probability p over N clock intervals, and A_i, $(0, 1)$, is the value of the logic level at the ith clock pulse, then the short-time averaging technique can be described by the equation

$$\hat{p}_N = \frac{1}{N} \sum_{i=1}^{N-1} A_i + A_N$$

i.e.

$$\hat{p}_N = \hat{p}_{N-1} + \frac{A_N - A_0}{N} \tag{2.16}$$

The smaller the sample size N the more effect the new value of A_i has on the estimate. When the average is taken over a short time interval the estimated probability, $\hat{p}_N$, responds quickly to change and is able to accommodate signals with a high harmonic content. When the value of N is increased the accuracy improves but the bandwidth is restricted.

The disadvantage of using this technique for filtering lies in the requirement to store all the logic levels present in the previous N clock intervals. However, since only the first and last levels are used in the calculation at any one time, a serial in/serial out shift register of length N can be chosen as the storage medium. Using these registers a circuit can be designed to implement a Moving Average, as in Fig. 2.15. For standard error of 1%, the length of the register must be more than 2500 stages, and for a 5% error the number drops to 100 stages.[32]

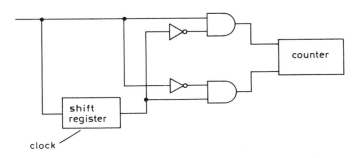

Fig. 2.15.   *Circuit for moving average*

So far we have only considered one form of averaging. However, any procedure which takes N observations of data and then forms a summation of the weighted readings, can be said to compute an average of the data. This is described by the equation

$$S_t = a_0 A_N + a_1 A_{N-1} + a_2 A_{N-2} + \ldots + a_{N-1} A_1$$

where S_t is the computed average at time t, i.e.

$$S_t = \sum_{i=0}^{N-1} a_i A_{N-1} \qquad (2.17)$$

For the average to be unbiased

$$\sum_{i=0}^{N-1} a_i = 1 \qquad (2.18)$$

If the subscript i is taken as referring to clock intervals in a digital system, the weighting coefficients can be plotted on a graph against time, and the relative importance which the average assigns to consecutive readings can be readily assessed. When all the coefficients, a_i, are made equal, eq. 2.17 describes a moving average:

$$\sum_{i=0}^{N-1} a_i = Na = 1$$

i.e.

$$a_i = 1/N \text{ for all } i$$

Any combination of coefficients which obeys eq. 2.18 can be used to form an average, and various examples are shown in Fig. 2.16. If N is taken as extending to infinity, eq. 2.18 can be satisfied by making the coefficients a geometric sequence.

$$a_1 = \alpha a_0$$

$$a_2 = \alpha^2 a_0$$

$$a_{N-1} = \alpha^{N-1} a_0$$

The sum of a geometric sequence is given by

$$\sum_{i=1}^{N-1} a_i = a_0 \frac{1 - \alpha^N}{1 - \alpha}$$

If $\alpha < 1$

$$\mathcal{L}_{N \to \infty} \left\{ \sum_{i=1}^{N-1} a_i \right\} = \frac{a_0}{1 - \alpha}$$

By letting $a_0 = 1 - \alpha$, eq. 2.18 is satisfied, and the average, S_t, is given by

$$S_t = (1 - \alpha)A_N + \alpha(1 - \alpha)A_{N-1} + \alpha^2(1 - \alpha)A_{N-2} + \ldots \qquad (2.19)$$

This method is known as exponential averaging or *Exponential Smoothing*, and its graph is drawn in Fig. 2.16d.[33]

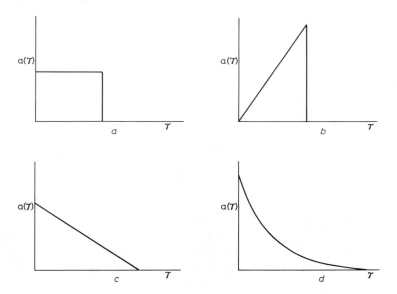

Fig. 2.16.   *Examples of weighting coefficients that sum to unity*

The equation which describes exponential smoothing can be rewritten as

$$S_t = (1 - \alpha)A_t + \alpha\{(1 - \alpha)A_{t-1} + \alpha(1 - \alpha)A_{t-2} + \ldots\}$$

$$= (1 - \alpha)A_t + \alpha \cdot S_{t-1}$$

This equation states that the new estimate is equal to the previous estimate plus a

correction term multiplied by $(1 - \alpha)$. The correction term is the difference
between the latest value of the input, A_t, and the previously estimated value i.e.

$$S_t = S_{t-1} + (1 - \alpha)(A_t - S_{t-1}) \tag{2.20}$$

The effect that the correction term has on the value of S_t depends directly on the
value of α. With $\alpha = 1$, the estimate will be unchanged by the new information, and
for $\alpha = 0$, the estimate will simply take on the present value of the data. If a signal
with a high noise content is to be averaged, the estimate, S_t, can be made insensitive
to random fluctuations by choosing α to be close to zero. If, on the other hand, the
signal has a fairly low noise level, a high value of α can be used, and any change in
the input signal will be quickly reflected in the average output. As with the moving-
average method there is a trade-off between noise rejection and bandwidth.

The digital circuitry for implementing exponential smoothing is shown in Fig.
2.17. At any given time instant t the previous estimate, S_{t-1}, is stored by the
counter in a binary number form. The number is transformed into a digital
sequence compatible with the input A_t and fed back to the input to act as a count-
down signal. Equilibrium is established when count-up and count-down rates equal-
ise. The value of α is governed by the length of the counter, being $(1 - 1/N)$ for an
N-state counter.

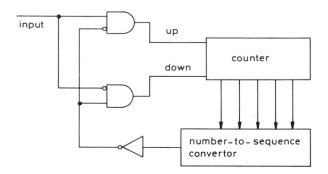

Fig. 2.17. *Exponential smoothing circuit*

It is informative at this stage to compare the results obtained for exponential
smoothing with the performance of a low-pass analogue filter. The voltage across a
capacitor and a resistor connected in parallel depends on the rate of charging the
capacitor. If the input to the parallel combination is a pulsed stochastic sequence
then the voltage developed across the capacitor forms an estimate of the stochastic
sequences generating probability. Each pulse in the stochastic sequence charges the
capacitor by a fixed amount. The voltage on the capacitor can therefore only vary
up to a certain maximum rate, and this rate is governed by the time constant of the
circuit and the frequency of the master clock pulse. Owing to the inability of the
circuit to respond to high-frequency fluctuations, it serves to filter out any high
frequency variations in probability.

It may be easily shown that the output voltage across the capacitor is related to the stochastic input sequence by the equation

$$v_t = v_{t-\Delta t} + \left\{ 1 - \exp\left(-\frac{\Delta t}{RC} \right) \right\} (A_t V_0 - v_{t-\Delta t}) \qquad (2.21)$$

Here V_0 is the voltage corresponding to an ON logic level, and A_t is the value of the logic level (0 ro 1) at time t. Δt represents the width of a pulse. Dividing eq. 2.21 by V_0 and rearranging gives

$$\hat{p}_t = \left\{ \exp\left(-\frac{\Delta t}{RC} \right) \right\} \hat{p}_{t-1} + \left\{ 1 - \exp\left(-\frac{\Delta t}{RC} \right) \right\} A_t$$

where

$$\hat{p}_t = \frac{v_t}{V_0} \quad \text{and} \quad \hat{p}_{t-1} = \frac{v_{t-\Delta t}}{V_0}$$

Thus $\hat{p}_t = \alpha \hat{p}_{t-1} + (1-\alpha) A_t \qquad (2.22)$

where $\alpha = \exp\{- [\Delta t/(RC)]\}$. Eq. 2.22 is identical to eq. 2.20 derived for exponential smoothing and this illustrates the fact that exponential smoothing is equivalent to using a low-pass analogue filter to suppress the high-frequency components of a waveform.

The digital synthesis in logic form of exponential smoothing is shown in Fig. 2.18. This circuit gives an output which is a measure of the generating probability of the stochastic input sequence. The circuit is sometimes called an ADDIE, which is an acronym for adaptive digital element.[4, 35] The operation of the ADDIE is dependent on both the stochastic input sequence and the probability of a feedback sequence obtained from the current state of the up-down counter. If the probability of the ADDIE being in state i at time t is $\pi_i(t)$, and the probability of changing from state i to state j at time t is $\pi_{ij}(t)$ then the probability of being in state j at time $t+1$ is

$$\pi_j(t+1) = \sum_{i=1}^{N} \pi_i(t)\pi_{ij}(t), \quad i = 1, 2, \ldots, j, \ldots, N$$

This shows that the operation of the ADDIE can be analysed as a nonstationary Markov process, with $\pi_{ij}(t)$ being a probability matrix with the rows summing to unity. The up/down counter's inability to jump states provides a restriction on the operation of the ADDIE. Thus, if the counter is in the initial state i, it can move to states $i-1$ or $i+1$, or it can remain in state i. It follows directly that

$$\pi_{i, i-1}(t) + \pi_{i, i}(t) + \pi_{i, i+1}(t) = 1$$

The inability of the ADDIE to jump states is equivalent to the restriction in the analogue low-pass filter case of the capacitor only being able to change its voltage by a certain fixed amount. It is this constraint which introduces a degree of 'inertia' in the system and provides the filtering action to stochastic sequences.

The counter of the ADDIE will count up if both inputs are ON and down if both inputs are OFF. When the inputs are different no change of state will occur. Let the probability of the input sequence being ON at any clock interval be p and the probability of it being OFF be q. The ADDIE is a chain-structured automaton with $N + 1$ ordered states $0, 1, 2, \ldots, N$. It may be shown that if the input signal is a stationary Bernoulli sequence, the states of the ADDIE will randomly fluctuate about a mean value given by [34, 35]

$$M = Np \tag{2.23}$$

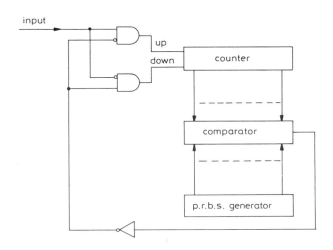

Fig. 2.18.   *Noise ADDIE*

The mean value will form an unbiased estimate of the input sequences generating probability and the distribution function of the ADDIE states will be binomial as given by

$$p_n = p^n q^{N-n} \binom{N}{n} \tag{2.24}$$

Eqs. 2.23 and 2.24 describe the steady-state characteristics of the ADDIE. These equations may be used to derive an expression for the accuracy of the ADDIE in its measurement of a given stochastic sequences generating probability. The statistics of the input stochastic sequence will not always be constant, and to assess the ADDIE's capability of dealing with nonstationary sequences it is necessary to determine its response to a step input.

Let $n(t)$ be the state of the ADDIE after t steps. It may be shown that the expected state of the ADDIE after t steps is given by [35]

$$\text{Exp}\{n(t)\} = N_p - \{N_p - n(0)\} \left(1 - \frac{1}{N}\right)^t \tag{2.25}$$

Eq. 2.25 indicates that the ADDIE will approach a new level along an exponential path. A time will be reached at which the inherent random fluctuations of the ADDIE states will be greater than the term

$$\{Np - n(0)\} \left(1 - \frac{1}{N}\right)^{t}$$

At this time meaningful readings may be taken from the ADDIE. Unless t is very large a bias term will exist owing to the exponential approach, and this bias term may only be decreased by decreasing the number of ADDIE states N. Although decreasing the number of ADDIE states increases the speed of response of the ADDIE to change, the error owing to the random fluctuation of the ADDIE states will also be increased.

The equivalence between the operating characteristics of the ADDIE and a low-pass analogue filter may be illustrated by comparing the equations describing their transient response to a step input.

For the analogue circuit

$$v_t = V_0 \left[1 - \exp\left(\frac{-t'}{\tau}\right)\right] \tag{2.26}$$

where t' = time in seconds, τ = circuit time constant and $V_0 = v(t)$ as $t \to \infty$.

For the ADDIE from eq. 2.25 with $n(0) = 0$ we have

$$\text{Exp}\{n(t)\} = Np \left\{1 - \left(1 - \frac{1}{N}\right)^{t}\right\} \tag{2.27}$$

where t = number of steps = $t'f_c$. Direct comparison of eqs. 2.26 and 2.27 gives

$$\tau = -\left\{f_c 1_n \left(1 - \frac{1}{N}\right)\right\}^{-1}$$

Eq. 2.27 may thus be rewritten as

$$n(t) = Np \left\{1 - \exp\left[t'f_c 1_n \left(1 - \frac{1}{N}\right)\right]\right\} \tag{2.28}$$

By multiplication of the time-constant term $-\{f_c 1_n (1 - 1/N)\}^{-1}$, the ADDIE's transient response can be equated to that of the analogue filter. However, owing to the noise added by the feedback sequence, the ADDIE will be less accurate than the analogue filter. A simple calculation shows that the variance of the analogue filter is given by $(pq)/(2N - 1)$. The variance of the ADDIE is obtained directly from eq. 2.24 as $(pq)/N$. Thus for the same transient response characteristics the analogue filter is $2 - 1/N$ times as accurate as the ADDIE structure.

The frequency limitations of stochastic computing systems may be conveniently stated in terms of the 3 dB point of the ADDIE. The time constant τ of the ADDIE obtained from eq. 2.28 is

$$\tau = -\left\{f_c 1_n \left(1 - \frac{1}{N}\right)\right\}^{-1}$$

Thus the 3 dB frequency is

$$W_{3\,dB} = -f_c l_n \left(1 - \frac{1}{N}\right) \tag{2.29}$$

If the error ϵ of the ADDIE is assumed to be the normalised standard deviation of the state distribution function, then

$$\epsilon = \left(\frac{pq}{N}\right)^{1/2}$$

Clearly the maximum error occurs when $p = q = 0.5$. Thus

$$\epsilon_{max} = 0.5(N)^{-1/2} \tag{2.30}$$

The bandwidth may be related to the error by substituting eq. 2.29 into eq. 2.30 to give

$$\epsilon_{max} = 0.5\left\{1 - \exp{-\frac{W_{3\,db}}{f_c}}\right\} \tag{2.31}$$

Fig. 2.19 shows a graph of error ϵ as a function of bandwidth normalised to clock frequency. Clearly the only method to reduce the slope of the graph to provide improved accuracy for a given bandwidth is to increase the value of the clock frequency f_c.

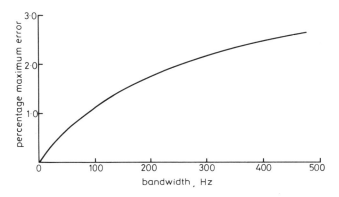

Fig. 2.19. *Percentage maximum error as function of bandwidth*

All the analysis so far has considered a noisy ADDIE structure with a feedback signal consisting of a Bernoulli sequence. If, alternatively, a deterministic feedback signal is used, the ADDIE will still form an estimate of the input sequence's generating probability. The implicit assumption in stochastic computation of statistical independence between signals is still applicable even if one of the signals is deterministic. The deterministic feedback signal may be conveniently generated using a binary rate multiplier (b.r.m.) in the feedback loop as shown in Fig. 2.20. To differentiate this structure from the noise ADDIE, we shall refer to it as a b.r.m. ADDIE.[35, 36] The output frequency f_b of the b.r.m. is determined by the current state of the up/down counter and the clock frequency f_c. The generating probability of the feedback sequence is f_b/f_c. Owing to the lack of random fluctuations in the feedback sequence the distribution function of the ADDIE states will be modified. In fact, as will be shown subsequently, the standard deviation is decreased compared with the noise ADDIE and this provides improved accuracy without reducing the bandwidth of the interface.

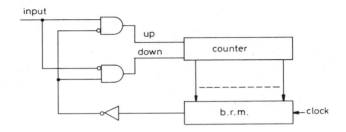

Fig. 2.20.   *Binary rate multiplier ADDIE*

The operation of the b.r.m. ADDIE can be classed as a nonMarkovian process and hence theoretical analysis is extremely difficult. However, it can be shown that a reduction in the standard deviation is to be expected and an expression which describes the extent of the reduction is developed below.

If the noise ADDIE is in a steady-state condition the variance of the distribution function is dependent on the variance of the count-up/count-down signals to the up/down counter. If the probability of an input pulse is p, then the probability of a feedback pulse is $(1-p)$ owing to the invertor in the feedback loop. The standard deviation of the count-up sequence, σ_n, for the noise ADDIE is thus

$$\sigma_n = [Np(1-p)]\,[1-p(1-p)]^{1/2} \qquad (2.32)$$

Similarly for the b.r.m. ADDIE, the standard deviation of the count-up sequence σ_b is given by

$$\sigma_b = \{(1-p)Np(1-p)\}^{1/2} \qquad (2.33)$$

Eq. 2.32 represents the standard deviation of a sequence with probability $p(1-p)$ computed over N clock intervals. Eq. 2.33 represents the standard deviation of a sequence with probability p computed over a sample size of $(1-p)N$ clock intervals. The reduction in standard deviation, as expressed by a reduction factor R_u, is obtained directly from eqs. 2.32 and 2.33 as

$$R_u = \frac{\sigma_n}{\sigma_b} = 1 + \frac{p^2}{1-p}$$

Thus R_u is maximum when $p=1$ and decreases to unity as p tends to 0. Similarly, for the count-down line the reduction ratio R_d may be calculated as

$$R_d = \frac{\sigma_n}{\sigma_b} = \frac{1-p(1-p)}{p}$$

This reduction ratio has a maximum at $p = 0$ and decreases to unity as p tends to unity. The resultant effect of both reduction ratios has been investigated experimentally for a range of input probabilities.[35]

Typical distribution curves for the noise ADDIE shown in Fig. 2.21 approximate well to the binomial distribution. This is consistent with the theoretical predictions of eq. 2.24. As predicted for the b.r.m. ADDIE, for a given input probability a decrease in variance occurs. The effect is most pronounced with input probability in the region of 0·5. It may be observed from Fig. 2.21 that the distribution curves for the b.r.m. ADDIE are asymmetrical and a sharp discontinuity occurs at the mean value. This discontinuity is a characteristic of the b.r.m. It is due to a change of phase of the sequence when the b.r.m. changes from a 'one-all-zero' state to a 'zero-all-one' state. The discontinuity occurs at probabilities of $1 - 1/(2^n)$ and $1/(2^n)$ where n is an integer.

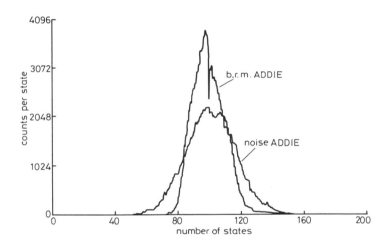

Fig. 2.21.   *ADDIE distribution curves*

The responses of both types of ADDIE to a range of step inputs are shown in Fig. 2.22. These curves were obtained in real time using a 12-bit digital-to-analogue converter and a graph plotter. It may be observed that the response of the b.r.m. ADDIE is almost identical to that of the noise ADDIE. The curves exhibit a close correspondence to the theoretically predicted curve obtained from eq. 2.27. According to eq. 2.27 the step response of the ADDIE is related to the total number of states. This was verified experimentally using the noise ADDIE and the results are shown in Fig. 2.23.

The frequency response of both types of ADDIE was measured for various clock frequencies and the results are shown in Fig. 2.24. As might be expected from the transient response results, little difference exists between the frequency response for both ADDIE's and the results show a close correspondence to the theoretical predicted curve obtained from eq. 2.29.

It should be noted that the moving and exponential averages are not the only possible alternatives for dealing with the averaging problem, and indeed do not

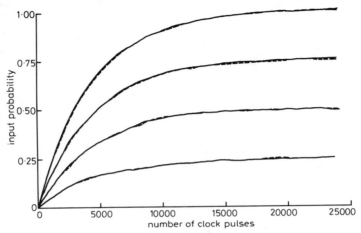

Fig. 2.22.   *ADDIE step response characteristics*

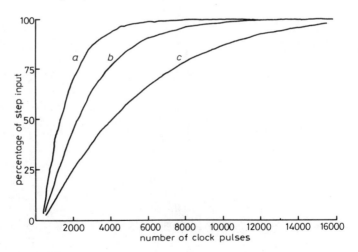

Fig. 2.23.   *Noise ADDIE step response as a function of number of states*
  a 10-bit
  b 11-bit
  c 12-bit

necessarily represent the optimum solution. They are of particular interest because their weighting operation can be easily duplicated using digital circuitry, and the circuitry, moreover, computes synchronously with the incoming sequence. It is possible to show theoretically that for stationary stochastic sequences exponential and moving average algorithms are equally viable, but for nonstationary sequences the moving average does possess an advantage because of its symmetrical weighting function.[37]

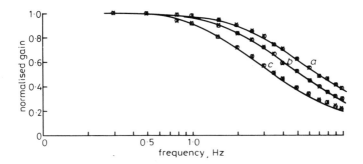

Fig. 2.24. *ADDIE frequency response characteristics*

The random numbers generated in a time stochastic processor need not pass all the conventional random number tests. Provided the number sequences are uniformly distributed and statistically independent, the probability measures will be unbiased and the fundamental theory of stochastic computing will apply. Apart from operator module modifications, any innovation to improve accuracy must be directed towards manipulating the properties of the individual number sequences. The critical element in sensing any improvement in accuracy is the output interface. The mean value of its estimate will measure any bias, and the variance will reflect any change in accuracy. To obtain a measure of the mean and variance, the states of the output interface counters are sampled at regular intervals and the average R of n random variables R_i is given by

$$R = \frac{1}{n} \sum_{i=1}^{n} R_i$$

The variance is

$$\text{var}(R) = \frac{1}{n^2} \sum_{i=1}^{n} \text{var}(R_i) + \frac{2}{n^2} \sum_{i \neq j}^{n} \text{cov}(R_1 R_j) \tag{2.34}$$

It should be noted that output interface circuits include a counter which can only change by one state per clock period. The covariance term in this case will be positive and cause an increase in the variance of the estimated mean. In order to counteract this effect the input sequence of logic levels should be generated with negative correlation.[18] For this condition, the covariance term of eq. 2.34 decreases and the accuracy of the estimate improves. Physically negative correlation implies an ordering of the logic levels to reduce the probability of long sequences of consecutive 1's and 0's. It is precisely these sequences which cause the counter to make extensive excursions away from the mean state, and thereby increase the measured variance.

Since any rearranging of the logic levels would require complex circuitry, the negative correlation is best derived from the sequences generation process. For the random sequence to be correlated, some deterministic operation must be introduced. If a Bernoulli sequence is passed down a shift register the random binary numbers formed by the stages will exhibit marked positive correlation. By inverting the outputs from even numbered stages as shown in Fig. 2.25*a*, the correlation may

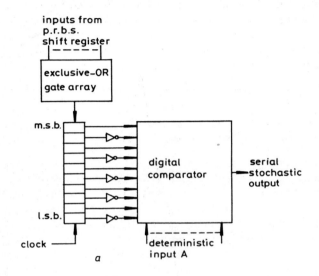

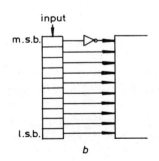

Fig. 2.25. *Negative correlation schemes*

be made negative.[18] An arithmetic operation will occur at every clock pulse which is intrinsic to the shift of data down the shift register. The number held in the shift register will be divided by two, which forms the most significant bit. An alternative circuit arrangement is to invert the output of the most significant bit and leave the other stages with direct outputs, as shown in Fig. 2.25*b*.

Fig. 2.26 shows typical experimental results obtained using negatively correlated sequences as inputs in a stochastic processing system. The two curves in Fig. 2.26 show the distribution of states obtained with a noncorrelated input probability. As expected an improvement in the variance of the output interface is obtained. Experimentally, little difference has been found between the two circuits of Fig. 2.25, but the circuit of Fig. 2.25*b* would be preferred because of its simplicity.

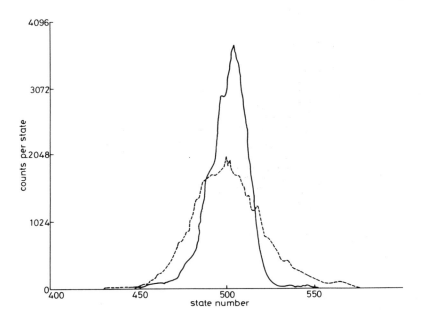

Fig. 2.26.   *Reduction in variance obtained with negative correlation*

The use of correlated bits has been further investigated and shown to decrease the nonlinearity of the random number distribution function. Under a Markovian assumption optimal permutations have been derived in graphical form.[38, 39] It is of interest to note that the process of stochastic conversion via an input interface may be considered to be a special case of random quantisation or *random reference quantisation.*[40–42] Such a generalisation permits the extension of the stochastic representation to an arbitrary number of bits.

Chapter 3

Applications of
stochastic processing

In this Chapter we consider some examples of stochastic processing machine design. Amongst the processors described are RASCEL, an early time stochastic array computer which first showed the practical flexibility of cascaded stochastic operations. The implementation of a general image transformer, TRANSFORMATRIX, demonstrates the exploitation of the enormous parallism afforded with simple arithmetic units. A variable topology time stochastic processor, APE, shows how one can achieve autonomous stochastic processing elements interconnected by radio-frequency channels. The design of a large general purpose time stochastic processor, DISCO, illustrates how flexibility of system configuration may be obtained using a supervisory minicomputer with real-time input/output. In the area of bundle processing we show that both failsoft (BUM) and failsafe (SABUMA) processor design is possible.

As one would intuitively expect, the most powerful application areas of stochastic processors are those involving the direct processing of probablistic variables. In particular we demonstrate in Chapter 4 that time stochastic processing methods provide an optimum synthesis technique for stochastic learning automata systems. Although for the simulation of deterministic ordinary differential equations stochastic methods are not competitive with either digital differential analysers or spatial arrays of microprocessors[43, 44] (a possible exception to this rule is noisy environments), there still exists a restricted class of deterministic algorithms for which stochastic implementations are attractive. One such class of problem concerns dynamic steepest ascent/descent computations. We give examples of experimental results including linear equations, matrix inversion and linear programming.

In the area of stochastic process simulation we consider Markovian dynamic systems and the Monte-Carlo solution of partial differential equations. Stochastic processors permit the high-speed computation of transient characteristics of Markov models and fast simulation of both stationary and time-dependent partial differential equations with arbitrary boundary conditions. Time stochastic processing is a potentially powerful tool in the area of neurophysiological modelling and we mention research work in this promising field.

The Chapter concludes with a discussion of reported applications in the fields of instrumentation and industrial process control. Of particular interest is the design of transducers for the direct production of stochastic sequences, stochastic digital/ analogue conversion and the stochastic implementation of signal correlators and spectrum analysers.

3.1 Examples of stochastic machines

The system POSTCOMP (*Po*rtable *St*ochastic *Comp*uter) was one of the first attempts to realise the principles of time stochastic processing in hardware form.[45] A 'signed absolute value' number system was used to map a number onto the probability of appearance of a pulse in a given time slot, its sign carried separately by an auxiliary wire. A subsequent design RASCEL (*R*egular *A*rray of *S*tochastic *Com*puting *El*ements) used a much more sophisticated number representation than POSTCOMP.[46] A two-wire (numerator wire, denominator wire) system was used to facilitate division, such that numbers between 0 and 1 had a machine representation between 0 and 0·5 while those between -1 and 0 were represented by probabilities between 0·5 and 1. Fig. 3.1 shows a photograph of RASCEL. The triangular structure at the top of the machine represents a tree arrangement of successive layers of stochastic processing elements (each capable of the four fundamental operations). Besides proving that stochastic elements could be cascaded to considerable depth, the work on RASCEL highlighted the attrition problem. This problem is similar to the fact that in a fixed-point computer, successive multiplications make the number processed (and therefore its accuracy) smaller and smaller. Different methods of circumventing this difficulty (by random duplication) were implemented.

Fig. 3.2 shows the fundamental idea of the machine TRANSFORMATRIX.[47-50] An $n \times n$ input matrix of points in the ij-plane is used, the intensity in (i, j) being represented by x_{ij}. Using n^2 coefficient matrices b_{klij} $(k = 1 \ldots n, l = \ldots n)$, the machine forms online the most general linear transform given by

$$y_{kl} = \sum_{i,j} b_{klij} x_{ij}$$

and displays y_{kl} on an $n \times n$ output matrix in (k, l). The interesting point is that stochastic arithmetic units permit the realisation of the case $n = 32$ in hardware. Thus TRANSFORMATRIX has 1024 parallel arithmetic units which form $\sum_{i,j} b_{klij} x_{ij}$ at once, and within less than 30 μs. This means that the output (1024 points) is refreshed every 1/30 of a second and is flicker-free. It should be noted that improved circuits (e.g. a 300 MHz clock instead of a 10 MHz clock and 1 s. refresh cycle) would permit a 10^6 point display.

The general linear transform encompasses the operations of translation, rotation and magnification, Fourier transformation, mask correlation for pattern recognition and general convolutions. It is therefore a very powerful technique. As shown in the schematic diagram of Fig. 3.3. TRANSFORMATRIX was one of the first stochastic machines to use pseudorandom noise to encode the outputs of phototransistors in in the ij plane. The machine uses online calculation of the coefficients b_{klij}, assembled in 1024 matrices $b_{1,1,ij} \ldots b_{32,32,ij}$. The calculation of the b_{klij} coefficients in the Fourier transform case is possible because of their periodicity. An overall photograph of TRANSFORMATRIX is shown in Fig. 3.4.

Fig. 3.1.   *Photograph of RASCEL*

An autonomous processing element system (APE) is a design based on an array of stochastic computers (each capable of addition, subtraction, multiplication, division, integration, differentiation and storage) communicating with each other over radio frequency channels and powered by light or microwave energy.[7, 51] The objectives of the system are to produce an array of integrated circuit chips which may be connected (under the control of an operator) in a desired topology; and to prove the feasibility of satellite computers which can be reconfigured after some form of failure. A schematic diagram of the APE design is shown in Fig. 3.5. Initial setup of the system is accomplished through a controller, which can send out directives on any of n (in this case $n = 14$, but $n = 1000$ is quite feasible) channels. Receiver B in Fig. 3.5 on detecting a directive (f.m. at frequency ν_k), produces a priority interrupt and sends the instruction signals to the function decoder. The instruction contains not only the type of operation to be performed, but also the a.m. frequencies on which the two input numbers are to be received. It should be noted that each element is uniquely characterised by the output frequency ν_k of its a.m. output. Duty cycle encoding makes all these outputs rise and fall in synchronism, the initiation signal being furnished by a clock broadcasting at a systems frequency ν_c. Since an element is named after ν_k, the parallelism of several elements corresponding to ν_k extends to the operation performed and the a.m. frequencies to which the inputs are tuned. It is trivial (by reassigning ordinal number k) to operate in the absence of a given k. Input transducers are incorporated which send their outputs to a subset of APE's, which in turn feed other APE's. The final output is made available to the controller which can remotely check all intermediate numbers and the functioning of any given APE.

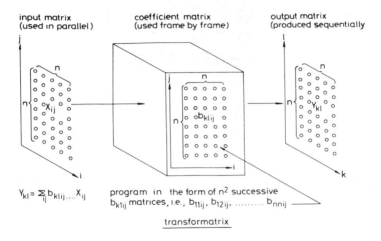

Fig. 3.2.   *TRANSFORMATRIX principle*

A general purpose digital stochastic computer DISCO shown in Fig. 3.6 was designed to be automatically programmed from a visual display unit.[52] This machine has been used to investigate a range of applications, including matrix operations, linear programming and process simulation. Instructions typed in are interpreted by a minicomputer which programs the stochastic machine and acts as an interface to the programmer. Thirty-four slots are available to provide any arithmetic operation, as dictated by the operator, and certain fixed slots provide inversion, multiplication and input interfaces. The system requires the use of the automatic patching system to interconnect any of 96 input nodes to 64 output nodes. It should be noted that because of the high frequencies involved (typical clock rates

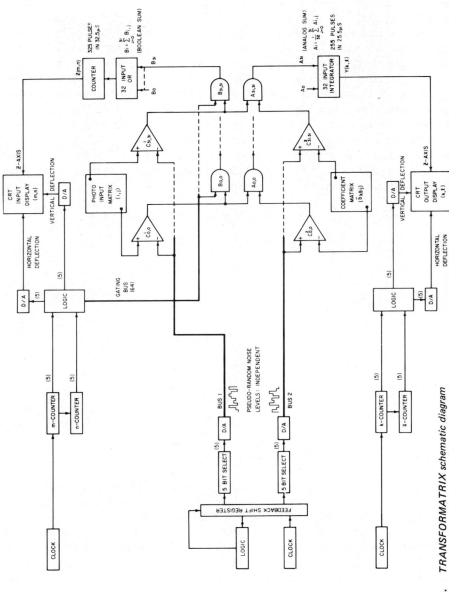

Fig. 3.3.   *TRANSFORMATRIX schematic diagram*

Fig. 3.4. *Photograph of TRANSFORMATRIX*

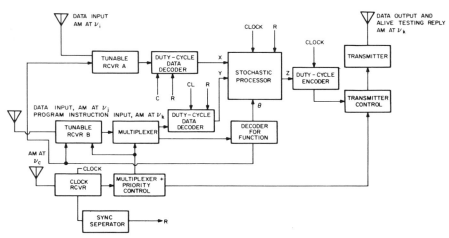

Fig. 3.5. *Schematic diagram of APE*

are from 1 MHz to 10 MHz) a manual patching system is not practicable. The inter-connection problem is solved using a multiplexing system based on data selectors. For any application, the network configuration established by the data selectors is dictated by a 6-bit code contained in a serial in/parallel output shift register, loaded from the minicomputer on command from a keyboard.

Fig. 3.6.   *Photograph of DISCO*

The bundle processing techniques discussed in Chapter 1 map (a finite number of) time slots onto the wires of a bundle. Fig. 3.7 shows a photograph of BUM (*Bundle Machine*).[8] This system is rather like the early POSTCOMP structure except that computations are failsoft in that small numbers of wires and transistors may be removed without materially influencing the end results. A subsequent design SABUMA (*Safe Bundle Machine*) shown in Fig. 3.8 extended the failsoft feature to provide nearly failsafe operation.[53] Thus while BUM can tolerate only a small number of failures, SABUMA survives as long as about 10% of all wires and transistors are operational. The fundamental principle of SABUMA was discussed in Chapter 1. If n_1 wires (out of N) are energised in a 'numerator bundle' and d_1 wires (out of N) are energised in a 'denominator bundle', the ratio $x_1 = n_1/d_1$ will remain constant when the probability of safe transmission is the same for both bundles. It is primordial to prove that arithmetic may be accomplished with ratios of bundles in a distributed arithmetic unit, that is, one in which no one element is critical.

Fig. 3.7. *Photograph of BUM*

As shown in the schematic diagram of Fig. 3.9 the *ergodic processor* ERGODIC is composed of three main units: the generators, the arithmetic unit and a processor.[9] A generator is simply a circular 64-bit shift register. Clearly, this method of generation of the signals for the bundle provides the ergodic properties automatically. The number of 1's and 0's loaded into the shift register determines the specific number carried by the bundle, both as a time average and as a cross-sectional average. As shown in Fig. 3.9 the arithmetic unit consists of randomisers and 74 arithmetic subunits. The function of the randomisers is to ensure that the two ergodic bundles are disjoint from each other; that is, if one wire in a bundle is

Fig. 3.8.   *Photograph of SABUMA*

energised, the corresponding wire from the other bundle is not energised and vice versa. Note that this property is only needed for addition and subtraction. The outputs of each arithmetic subunit are collected to form the resultant bundle. A photograph of the ERGODIC processor is shown in Fig. 3.10.

3.2 Deterministic simulations

We have demonstrated in Chapter 2 that all of the conventional analogue computer elements may be realised using stochastic processing techniques. It follows directly that such structures may be used to simulate deterministic ordinary differential equations and active filters.[54] A typical example is given in Fig. 3.11 which shows the stochastic implementation of a second order differential equation. Essentially

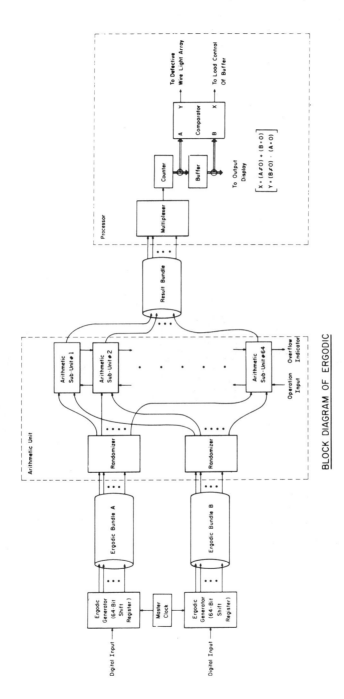

BLOCK DIAGRAM OF ERGODIC

$$\begin{bmatrix} x = (A \neq O) + (B = O) \\ Y = (B \neq O) \cdot (A = O) \end{bmatrix}$$

Fig. 3.9. *Schematic diagram of ERGODIC*

the results obtained are as expected from standard analogue realisations with super-imposed random noise. Recent work has considered a variety of possible filter structures and the synthesis of stochastic phase-locked-loops.[55, 56] These systems all have significantly high noise immunity, but it should be noted that advances in integrated circuit design mean that their applicability can no longer be justified on the basis of economic circuit fabrication.

Fig. 3.10.   *Photograph of ERGODIC*

A promising area in deterministic simulations is the application of gradient techniques. Such methods were first suggested in the context of stochastic processing for implicit function generation and linear system identification.[15, 57] Subsequently it has been shown that gradient algorithms can be used to provide the stochastic processing realisation of a variety of problems including linear equations, matrix inversion and linear programming.[58, 59] The main potential advantage of such an approach is to permit fast computations for the large scale real-time control of industrial processes in noisy environments. As an example of the computation of linear equations and matrix inversion using time stochastic processing consider the set of simultaneous equations, in matrix form, written as

$$Ax = b \tag{3.1}$$

or

$$\sum_{k=1}^{n} a_{ik}x_k = b_i$$

Here x and b are n-vectors and A is an $n \times n$ matrix. The problem may be retranslated as an optimisation problem by defining an error vector e such that

$$Ax - b = e$$

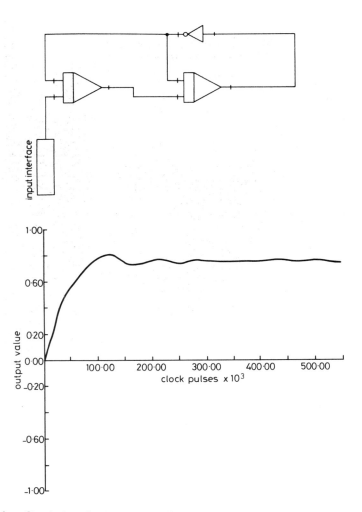

Fig. 3.11. *Simulation of second-order differential equation*

Obviously, $e = 0$ when the vector x satisfies the origin equation. A suitable criterion function is:[60]

$$f = \frac{1}{2} e^T e = \frac{1}{2} \sum_{i=1}^{n} e_i^2$$

Using steepest descent, we have

$$\dot{x} = -K\nabla f \quad \text{with } K > 0$$

and hence

$$\dot{x}_j = - K \sum_{i=1}^{n} a_{ij} \left\{ \sum_{i=1}^{n} a_{ik} x_k - b_i \right\}$$

n equations of this form must be implemented on the stochastic machine. The basic schematic of the machine solution for the generalised case is shown in Fig. 3.12.

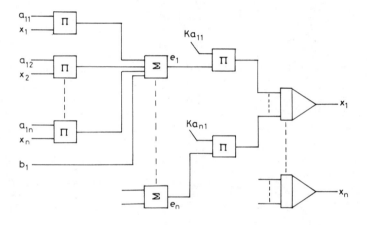

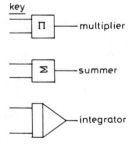

key

П ——— multiplier

Σ ——— summer

▷ ——— integrator

Fig. 3.12.   *Schematic diagram for linear equation simulation*

It is of considerable interest, particularly from the online control aspect that the structure shown in Fig. 3.12 may also be used for fast matrix inversion. Eq. 3.1 may be rewritten as

$$x = A^{-1}b \qquad (3.2)$$

Suppose we wish to invert the matrix A, and

$$A^{-1} = \begin{bmatrix} C_{11} & C_{12} \dots C_{1n} \\ \\ C_{21} & C_{22} \\ \vdots & \vdots \\ C_{n1} & C_{n2} \dots C_{nn} \end{bmatrix}$$

Eq. 3.2 may be rewritten as

$$\begin{bmatrix} x_1 \\ x_2 \\ \vdots \\ x_n \end{bmatrix} = \begin{bmatrix} C_{11} & C_{12} & C_{1n} \\ C_{21} & C_{22} & C_{2n} \\ \vdots & & \vdots \\ C_{n1} & C_{n2} \dots C_{nn} \end{bmatrix} \begin{bmatrix} b_1 \\ b_2 \\ \vdots \\ b_n \end{bmatrix}$$

Since the original system solved for x, we need to make the vector b successively

$$\begin{bmatrix} 1 \\ 0 \\ \vdots \\ 0 \end{bmatrix}, \begin{bmatrix} 0 \\ 1 \\ \vdots \\ 0 \end{bmatrix}, \dots, \begin{bmatrix} 0 \\ 0 \\ \vdots \\ 1 \end{bmatrix}$$

to generate the columns of the inverted matrix automatically. This can easily be accomplished using an n-state counter.

As a practical example, consider the simple 3rd-order system described by

$$\begin{bmatrix} 2 & 1 & 1 \\ 1 & 2 & 1 \\ 1 & 1 & 2 \end{bmatrix} \cdot \begin{bmatrix} x_1 \\ x_2 \\ x_3 \end{bmatrix} = \begin{bmatrix} 1 \\ 2 \\ 3 \end{bmatrix}$$

The known solution of these equations is

$$x = \begin{bmatrix} -0.5 \\ +0.5 \\ +1.5 \end{bmatrix}$$

Fig. 3.13 shows the state trajectories obtained with a stochastic implementation. At a clock frequency of 10 MHz steady-state conditions are acheived in approximately 0·2 ms. The trajectories exhibit the deterministic behaviour associated with conventional analogue solutions, but with superimposed random variance. In the steady state, a trajectory, e.g. x_1, can be slightly to one side of the optimum, owing to the intrinsic difficulty of small error estimation. Despite this, in the steady-state system error is of the order of 1% full scale, which is more than adequate for many online control problems. In fact, at the expense of computation time, any desired accuracy may be obtained by increasing the integrator bit capacity.[58]

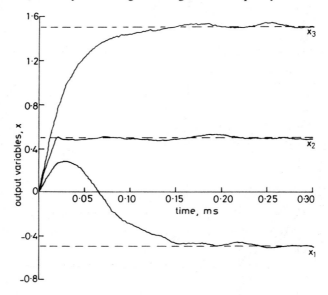

Fig. 3.13.   *Experimental results for linear equations*

We have previously noted that a slight modification of the circuit used for linear equation solution permits high-speed matrix inversion. Consider the problem of the inversion of the matrix A given by

$$A = \begin{bmatrix} 1\cdot0 & 0\cdot5 & 0\cdot5 \\ 0\cdot5 & 1\cdot0 & 0\cdot5 \\ 0\cdot5 & 0\cdot5 & 1\cdot0 \end{bmatrix}$$

the known solution is given by

$$A^{-1} = \begin{bmatrix} -1\cdot5 & -0\cdot5 & -0\cdot5 \\ -0\cdot5 & 1\cdot5 & -0\cdot5 \\ -0\cdot5 & -0\cdot5 & 1\cdot5 \end{bmatrix}$$

Fig. 3.14 shows the state trajectories obtained for the computation of the first column in the inverted matrix. The steady-state values are reached in approximately 0·2 ms, with errors again within 1% of full scale. Total matrix inversion in this case is obtained in less than 1 ms.

The important point to note is that, because of the parallel nature of the stochastic computation, the matrix inversion time is not significantly influenced by the size of matrix. In practice, some degradation inevitably occurs because attenuation factors introduced by summation elements necessitate the use of larger integrator bit capacities, with a corresponding increase in computation time for a given accuracy.

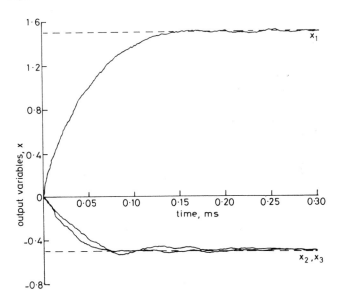

Fig. 3.14. *Experimental results for matrix inversion*

The above results on linear equations may be extended to linear programming algorithms. Linear programming is widely used as a basic technique for the control of complex industrial processes. For n variables and m constraints the linear programming problem may be stated as

$$\text{Minimise } F = \sum_{k=1}^{n} a_k x_k (x_k \geqslant 0)$$

subject to m restrictions of the form

$$\sum_{k=1}^{n} b_{mk} x_k \geqslant 1 \tag{3.3}$$

The problem may be solved using a steepest-descent path modified by the existence of the restricted regions in n-dimensional space.[61] The time variation of the variables x_i may be described by a vector equation of the form

$$\dot{x} = -K\nabla F + \sum_{j=1}^{m} \alpha_j \gamma_j \tag{3.4}$$

where

$$\alpha_j = \begin{cases} 0 \text{ if } \sum_{k=1}^{n} b_{mk}x_k \geqslant 1 \\ \\ 1 \text{ if } \sum_{k=1}^{n} b_{mk}x_k < 1 \end{cases}$$

and the vector

$$\gamma_j = b_{j1}i_1 + b_{j2}i_2 + \ldots + b_{jn}i_n$$

is normal to

$$\sum_{k=1}^{n} b_{mk}x_k = 1$$

Eqs. 3.3 and 3.4 may be implemented with stochastic modules, as shown in Fig. 3.15. The threshold switches used to sense the crossing of a constraint boundary are synthesised using the adaptive digital logic elements (ADDIES) described previously.

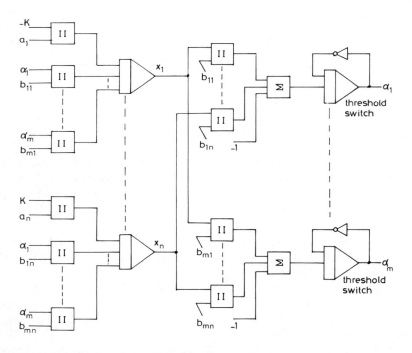

Fig. 3.15.   *Schematic diagram for linear programming*

To illustrate typical results obtained for linear programming, consider the simple problem:[59]

$$\text{maximise } Z = 0.5x_1 + x_2 + 0.333x_3$$

subject to the constraints

$$x_1 + 0\cdot666x_2 + 0\cdot166x_3 \leqslant 0\cdot666$$

$$0\cdot666x_1 + x_2 + 0\cdot666x_3 \leqslant 0\cdot666$$

$$x_1, x_2, x_3 \geqslant 0$$

The standard simplex method applied to this problem gives

$$Z_{max} = 0\cdot666 \text{ and } x_{opt} = \begin{bmatrix} 0 \\ 0\cdot666 \\ 0 \end{bmatrix}$$

Fig. 3.16 shows the state trajectories obtained for a stochastic implementation of the problem. Steady-state solutions are obtained in approximately 0·3 ms and error is less than 1% of full scale. It is of interest to note the discontinuities which occur

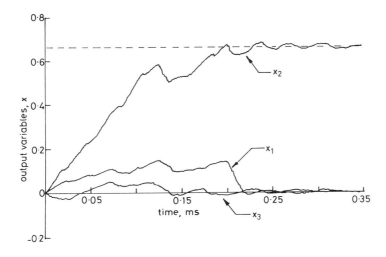

Fig. 3.16. *Experimental results for linear programming*

in the state trajectories at given instants. This 'zigzag' effect is caused by the state vector violating the constraint conditions. If the constraints are satisfied, the equation defining the motion in state space is given by eq. 3.4 with $\alpha_j = 0$. If the point attempts to cross a given constraint boundary, α_j changes from 0 to 1 and a component of velocity normal to and away from the constraint plane is imparted to the travelling point. This produces the characteristic 'zigzag' motion. The size of the discontinuity is somewhat amplified by the fact that the ADDIE's inertia produces a delay between the moving point crossing a boundary and the ADDIE's switching response. In practice, this effect can be largely reduced by using ADDIES of very much smaller bit capacity compared with the integrators. Obviously, some compro-

mise is involved, because of the usual conflict between accuracy and computation
time.

Considerable interest has been shown in the application of spatial arrays or nets
of stochastic processing elements. Such arrays have been shown to permit matrix
multiplication, space variable integration and approximate space-variable integral
transformation.[13] Fig. 3.17 shows a schematic diagram for a simple (2×2) array
matrix multiplier. The resultant product terms are given by

$$C_{mn} = \frac{1}{2} \sum_{i=1}^{2} a_{mk} b_{kn}$$

where the factor of $\frac{1}{2}$ appears due to the summation process. Such array structures
can be easily extended to higher dimensions.

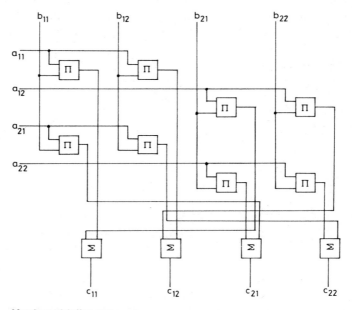

Fig. 3.17. *Matrix multiplier array*

Spatial arrays of special purpose stochastic processing elements have been suggested for the simulation of partial differential equations.[15, 62] The approach
adopted is similar to that of the conventional analogue computer solution. Consider
the linear autonomous first order system described by:

$$\frac{\partial U_N}{\partial t} + a_N \frac{\partial U_N}{\partial x} + \sum_{k=1}^{N} b_{Nk} U_k = 0$$

where $U_1 \ldots U_N$ are unknown functions of x and t; a_N and b_{Nk} are parameters
specified by the particular problem. The system may be solved by discretising the
space variable x into M intervals using the first-order approximation

$$\frac{\partial U_j}{\partial x} = \frac{1}{\Delta x}|U_j(x_i) - U_j(x_{i-1})|$$

where $x_{i-1} < x < x_i$ and $\Delta x = x_i - x_{i-1}$.

This solution leads to a system of MN differential equations described by

$$\frac{dU_j(x_i)}{dt} + \frac{a_j}{\Delta x}|U_j(x_i) - U_j(x_{i-1})| + \sum_{k=1}^{N} b_{jk}U_k(x_i) = 0 \qquad (3.5)$$

In terms of a stochastic array implementation each of the equations described by eq. 3.5 is solved by a specialised module of the form shown in Fig. 3.18. Networks of such modules can be used to solve a wide variety of partial differential equations in two dimensions for arbitrary shapes and boundary conditions.[62]

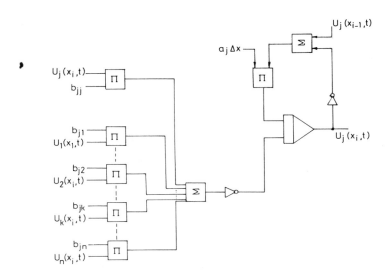

Fig. 3.18   *Specialised stochastic module*

Stochastic processing arrays have been proposed for the simulation of hydro-dynamic flow problems.[10] Essentially, the molecules of a liquid are mapped into stochastic sequences entering and leaving specialised arithmetic cells. For example, it is known that the lift of an airfoil is proportional to the speed vector over its surface. The mathematical simulation of this turbulent flow is extremely difficult because rapidly converging solutions are only known for cases in which there is a speed potential, and this excludes turbulent flow. Consider a 2-dimensional problem in which an incompressible liquid flows inside a given boundary, owing to a finite number of sources and sinks. Fig. 3.19 shows a possible stochastic array. The divergences D_{ij} may be positive or negative. Except at the source or sink points we must have div $v = 0$. This condition is simulated by arranging that the number of pulses entering a cell from any direction equals the number leaving the cell in the same time period. The fundamental idea is to iterate by observing the accumulation of pulses when the sum entering a cell is distributed according to a certain pattern and to modify the distribution for the next iteration.

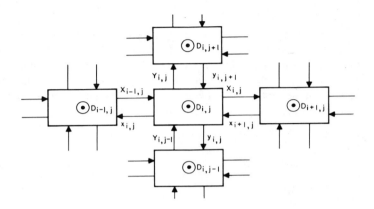

Fig. 3.19.   *Stochastic array for hydrodynamic simulation*
Stochastic sequences

$D_{i,j} \rightarrow$ divergence in cell (i, j) — usually zero
$X_{i,j} \rightarrow$ right-flow
$x_{i,j} \rightarrow$ left-flow
$Y_{i,j} \rightarrow$ up-flow $\Bigg\}$ out of cell (i, j)
$y_{i,j} \rightarrow$ down-flow
$\Sigma_{i,j}^{N} =$ sum of inflows
$S_{i,j}^{N} =$ sum of outflows $\Bigg\}$ in period N
$D_{i,j}^{N} =$ sum of divergences

3.3 Stochastic process simulation and modelling

Stochastic models are being used to an ever increasing extent to investigate phenomena that are essentially concerned with a random flow of events in time, especially those exhibiting such highly variable characteristics as birth, death, queueing, evolution, etc. One such stochastic model is that of Markov Chains which have been used extensively in the field of operations research for many years. While Markov Chains have been applied successfully to many areas in operational research, few high speed simulation models exist. To simulate Markov models using a conventional digital computer requires the use of many iterative procedures which are very slow, leading to considerable solution times.

A finite state Markov Chain may be described by a stochastic transition matrix P where

$$P = \begin{bmatrix} p_{11} & p_{12} \cdots p_{1n} \\ p_{21} & p_{22} & \vdots \\ \vdots & & \vdots \\ p_{n1} \cdots \cdots \cdots p_{nn} \end{bmatrix}$$

and p_{ij} represents the probability of a state transition from state i to state j.

And we define a state probability $\pi_i(n)$ as the probability that the system occupies state i after n transitions, if the state at $n = 0$ is known, then

$$\sum_{i=1}^{N} \pi_i(n) = 1$$

$$\pi_j(n + 1) = \sum_{i=1}^{N} \pi_i(n) P_{ij} \text{ for } n = 0, 1, 2 \ldots$$

If we define a row vector of state probabilities $\pi(n)$ with components $\pi_i(n)$ then

$$\pi(n + 1) = \pi(n)P$$

In general

$$\pi(n) = \pi(0)P^n \text{ for } n = 0, 1, 2 \ldots \tag{3.6}$$

Hence the probability that the system occupies each of its states after n moves, $\pi(n)$, is given by postmultiplying the initial state probability vector $\pi(0)$ by the nth power of the transition matrix P. The requirement is therefore for a high-speed simulation of eq. 3.6.

The requirements for a hardware Markov simulator are as follows:

(*a*) an n state sequential network with programmable inputs and state detection
(*b*) the facility to program the probabilities of the stochastic matrix
(*c*) a programmable counter which will determine the number of transitions n
(*d*) the ability to estimate the probability of being in any one state after n transition periods.

Figure 3.20 shows the general schematic diagram for a Markov Chain simulator.[63, 64]

The first requirement is met by using an n state sequential network consisting of flip-flops and appropriate combinational logic. This means that one change of state or transition will occur at every clock pulse, i.e. one matrix multiplication will be accomplished by each clock pulse. This fact is very significant from the aspect of computational speed. The components of the stochastic matrix are generated by stochastic comparators, each one giving a stochastic sequence with a variable probability of delivering an ON pulse. This means that for the duration of any clock period the stochastic matrix will be composed of 'ones' and 'zeros' with no intermediate values possible. Intermediate values are in fact represented by the probabilities of finding a one' in each matrix position. Therefore, because each row in the matrix must sum to unity, there can only be one 'one' in each row and probability transformers must be used to ensure that this is the case. Secondly, it is meaningless to have a nonzero value for any component of a row other than that row which corresponds to the state occupied at the time of the clock pulse. This would mean that the system would be required to vacate a state which it does not occupy at the time of the clock pulse. Because of these last two conditions only one of the transition probabilities can be 'one' at any time.

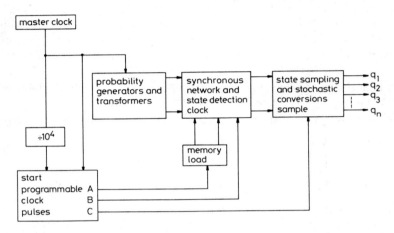

Fig. 3.20.   *Schematic diagram for Markov simulator*

As shown in Fig. 3.20 a programmable clock pulse generator is used to clock the synchronous network a specific number of times. State sampling and ADDIE circuits are used to estimate the probability of state occupancy after a particular number of clock pulses.

Fig. 3.21 shows typical experimental and theoretical results obtained for a four state Markov system. It should be noted that although for convenience the theoretical predictions are shown as continuous curves, the probability of state occupancy is of course only defined for integer values of clock pulses. The experimental results are in close agreement with the theoretical predictions and confirm the expected result that steady-state probabilities are reached within relatively few clock pulses. At a typical clock frequency of 1 MHz this means that steady state probabilities for Markov systems may be computed within a few microseconds. The important point to note is that, because of the parallel nature of the computations, this speed is not influenced by the size of the matrix involved.

Random-walk models are widely used in Monte-Carlo methods, especially in the solution of partial differential equations, multiple integrals and in the study of diffusion processes. Again, the alternative techniques, e.g. finite differences implemented on a conventional digital computer, involve considerable computing hardware and a complete solution over the entire region of interest is required to yield a single-point solution. Analogue-hybrid computer methods have been suggested and implemented,[65, 66] giving fast solution, but lacking some of the advantages of digital implementation.

The random-walk technique is based on the solution of the stochastic differential equations of motion of a point governed by n coupled first-order nonlinear equations, these being excited by n zero-mean white Gaussian noise signals. The solution of partial differential equations with specific boundary values is found by starting a random walk at the point of interest in, for example, a three-dimensional grid, and in the case of elliptic equations, continuing the walk until a boundary is reached, whereupon the boundary value corresponding to the crossing point is tallied and a new walk initiated. In the case of parabolic equations, the random walk continues for a specific time, i.e. number of steps, and if a boundary is crossed then the boundary value is tallied, otherwise the initial value corresponding to the starting point is tallied, and a new walk initiated. The solution corresponding to the starting point is the final tally divided by the number of walks.

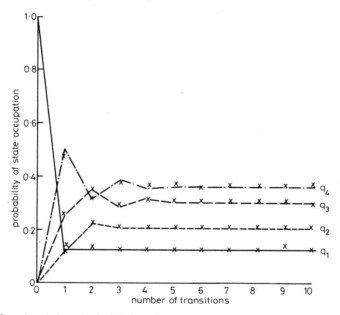

Fig. 3.21. *Experimental results for Markov chains*

It may be observed that a random walk is a special case of a Markov Chain with the restriction that it cannot 'jump' states. Considering a one-dimensional case, using the concept of the stochastic matrix described previously, only the transition probabilities $p_{n, n+1}$ and $p_{n, n-1}$ can exist, i.e. the particle can only move to an adjacent state. These transitional probabilities will be constant for all n, and other transitional probabilities will be zero. This may of course be extended to the multidimensional case.

Stochastic hardware simulation of random walks may be simply performed using stochastic integrators and appropriate control logic and memory. Fig. 3.22 shows a schematic diagram for a three-dimensional random-walk simulator. The stochastic integrators have programmable probabilities of counting up/down or remaining

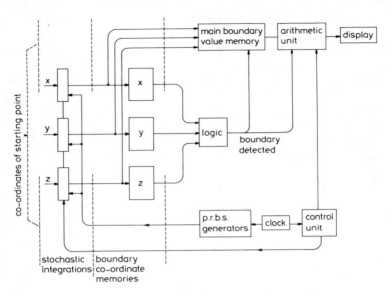

Fig. 3.22.   *Schematic diagram for random-walk simulator*

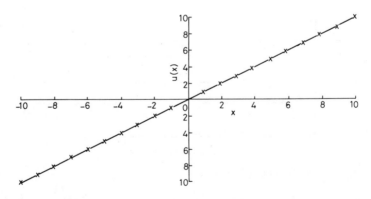

Fig. 3.23.   *Experimental results for Laplace's equation*

stationary. Programmable read-only memories are used to store arbitrary boundary conditions and control logic and arithmetic units permit the convenient simulation of elliptic or parabolic partial differential equations and selection of absorbing or reflecting boundaries. Figs. 3.23 and 3.24 show typical experimental results for the simple case of Laplace's and the Diffusion equation in one dimension. Typical solution times are less than 1 s for errors of the order of ±1%. For more complex problems, the stochastic hardware implementation of Monte-Carlo techniques has significant potential for providing fast economic computations for the synthesis of control algorithms for the on-line control of distributed systems,[67] and for the solution of partial differential equations with arbitrary boundary conditions in n-dimensions for both stationary and time dependent problems.

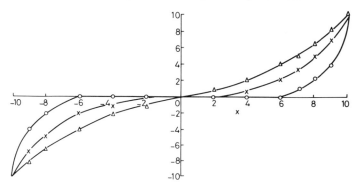

Fig. 3.24. *Experimental results for Diffusion equation*

Time stochastic processing possesses some similarities to the neural systems behaviour of the brain, and it has been suggested that stochastic networks can be applied to neural net modelling, and as a simulator for some cortical functions, in particular the cross-correlation processes of visual disparity and auditory-format separation.[13, 15] Simple stochastic processor models of sensory reception in the eye and the simulation of nerve net cells with application to pattern recognition with feature extraction have been reported.[15, 68] Recently, Fourier-Lattice models have been considered as a possible class of structures for brain process modelling. Stochastic techniques have been used to simulate such real-time tensor lattice networks.[69]

3.4 Instrumentation and control applications

It is not generally realised that because of the atomic or quantised nature of the universe, most information emanating from physical systems occurs in the form of a random sequence of pulse. A typical example would be a photocell in which each photon produces an output pulse. Of course, we generally only see the integral of the pulses because of their overlap and the limited resolution or bandwidth of the measuring equipment. The idea of using statistical properties is not new, e.g. fuel flow has been measured by Geiger counters and radiocative admixtures. It is thus of considerable interest to consider the possible design of stochastic transducers which permit the direct production of random sequences from physical variables without the usual input interface transformation problems prior to stochastic processing.

As an example, Fig. 3.25 shows a method of measuring the speed of a liquid

without the use of radioactive particles (carried by the flow) and Geiger counters.[10, 70] The principle is to generate vortices by an obstacle in the flow, and to
count these vortices by means of some form of pressure transducer (e.g. micro-
phone). The vortices occur at irregular intervals and the pressure transducer
measures a random process which is dependent on the average flow. Note that we
are interested in the fluctuating pressure measurement and not the average pressure.
The pressure transducer will output a stochastic sequence. Temperature transducers
based on molecular phenomena have also been investigated and a combination of
pressure and temperature transducers used in conjunction with stochastic processing to implement thermal flow measurements.[10, 70] In the context of fluid flow
it is of interest to note that a range of fluidic logic devices based on naturally
occuring random phenomena and stochastic processing have been proposed.[71, 72]

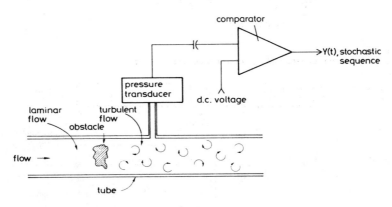

Fig. 3.25.   *Arrangement for flow transducer*

   A family of stochastic processing instrumentation systems including measuring
instruments, waveform analysers and correlators have been developed and offered
commercially.[73] Such instruments can process both d.c. and a.c. signals up to 1 MHz
frequency without regard to waveshape and ignoring either contaminated signal or
intrinsic instrumentation noise. For power and correlation measurements, stochastic processing techniques do not exhibit the gain range and linearity problems
associated with analogue circuits. Accuracy for the instruments is typically within
±1% of full scale for signals from 15 Hz to 1 MHz. Correlator systems can accomodate signals with crest factors up to 15 for frequencies from d.c. to 1 MHz. A
similar family of stochastic ergodic instruments has also been developed primarily
for educational and research purposes.[74]
   Also in the area of signal processing linear prediction filters have been proposed,
synthesised using stochastic techniques.[75] The main advantages of such structures
for slow applications is the simplicity of the arithmetical circuits and high-noise
immunity characteristics. (As discussed subsequently, burst processing is also a
promising technique for prediction filtering.[83]) In addition, the application of

one-bit random reference quantisation to correlation function estimation and Fourier Transform computation has been considered.[76] Correlation function and power spectral density on 512 points for wide band signals (several tens of MHz) is possible.

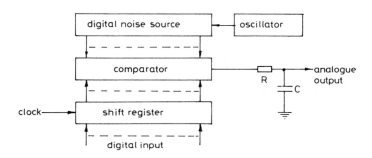

Fig. 3.26. *Principle for stochastic digital/analogue conversion*

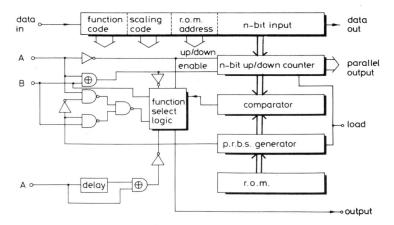

Fig. 3.27. *Universal stochastic module*

In the area of special-purpose instrumentation an electricity consumption meter based on stochastic techniques has been designed.[77] The system involves a logarithmic two-line bipolar stochastic conversion of the three-phase multiplexed input voltage and current signals, bipolar multiplication and nonlinear accumulation/ integration. It is found that the resolution and statistical error relative to the measured power value are constant over the entire operating range. Such a system is very suitable for large scale integrated circuit implementation and possesses a significant tolerance to power line noise and surges.

The first commercially available integrated circuit using stochastic techniques was the 10-bit digital/analogue converter shown schematically in Fig. 3.26.[78] This is readily identified as the standard digital/stochastic input interface introduced in Chapter 2 with the addition of a low-pass filter. The converter avoids the use of a resistor-ladder network and utilises the linear relationship which exists between the mean value of the stochastic output signal and the digital input required to be converted. A simple averaging process obtained by means of the low-pass filter produces an equivalent analogue voltage.

Reported applications of stochastic processing to industrial process control have included the stochastic control of thermal heating systems[79] and turbine speed control.[80] Electrical aircraft flight control (fly-by-wire systems) have been implemented using stochastic techniques.[81] The feasability of control applications would be enhanced by the possible development of some form of universal stochastic processing module (u.s.m.). Such a module could provide any interface or arithmetic operation as dictated by a programmed input control signal. One possible schematic for such a module is shown in Fig. 3.27. In this circuit, the scaling factor for the up/down counter, the ROM address, function select code and input information are all entered via a large serial shift register. It is envisaged that this data would be loaded under control of a supervisory microcomputer with the u.s.m.s connected in an iterative array to synthesise a particular control algorithm.

Chapter 4

Stochastic learning automata

There exist several areas in which the property of randomness is essential for computation.[85-87] Examples include the provision of random weights to ensure convergence of adaptive threshold logic,[15] the demonstration that a class of control problem exists soluble by two-state stochastic automata but insoluble for finite state deterministic automata[88] and the synthesis of stochastic learning automata.

In many process control problems, the characteristics of the process are fully known, and a complete mathematical description of the process and of the corresponding control strategy is possible. However, a large number of situations arise where uncertainties are present, either due to an incomplete mathematical model of the process, or due to operation in a random environment. Where the probablistic nature of these uncertainties is known, stochastic control theory can be applied, but in the case of higher-order uncertainties where the probablistic characteristics cannot be easily ascertained, it is only possible to gain sufficient knowledge of the process by 'on-line' observation. Herein lies the application area for stochastic learning automata. Essentially, stochastic learning automata provide a novel and attractive mode of solving a large class of problems involving uncertainties of a high order. Many problems of adaptive control, pattern recognition, filtering and identification can, under proper assumptions, be regarded as parameter optimisation problems. A learning automaton can be fruitfully applied to solve such problems, especially under noisy conditions when the *a priori* information is small. In fact, the possibility of using a stochastic automaton as a model for a learning controller provided the first motivation for studies in this area. For such problems, unlike stochastic approximation methods, the learning automaton has the desired flexibility not to get locked on to a local optimum, and this fact makes the automata approach particularly applicable to multimodal performance criteria systems.

This Chapter considers the theory and applications of stochastic learning automata. In particular we show that stochastic processing provides an optimum synthesis technique for such automata and permits the economic realisation of high-speed controllers for real-time system control.

4.1 Learning algorithms

Early work on learning automata models was developed in the context of mathematical psychology.[89-92] Recently, an excellent review paper and a special volume of papers devoted to learning automata have appeared.[93, 94] In general, a learning automaton may be defined as an automaton which interacts with a random environment in such a manner as to improve a specified overall performance by changing its action probabilities dependent on responses received from the environment.

A learning automaton may be described by a quintuple $\{x, \phi, a, F, G\}$, where x is the input set, ϕ is the set of internal states $\phi = \{\phi_1, \phi_2 \ldots \phi_s\}$ and a is the set of output states where $a = \{a_1, a_2 \ldots a_r\}$ and $(r \leqslant s)$. The state transition mapping F relates $(x, \phi) \rightarrow \phi$ and the output function G relates the set of internal states ϕ to the output states a, that is $G: \phi \rightarrow a$. If the elements of the matrices F and G are deterministic $(0, 1)$ the automaton is classified as *fixed structure deterministic*. Alternatively, the elements of F and G can be constant probabilities in which case the automaton is considered as *fixed structure stochastic*. In a variable structure automaton the state probability vector, $p(n)$, governing the choice of the state at each stage (n) is dependent on both its previous value and an updated version of the state transition matrix F. A reinforcement or updating algorithm is used to change F dependent on responses from the environment.

Fig. 4.1 shows a learning automaton interacting with an environment characterised by a penalty probability set c_i $(i = 1$ to $r)$, where c_i represents the probability that automaton action a_i is penalised by the environment. The actions of the automaton form the inputs to the environment and the responses of the environment act as inputs to the automaton. The input to the environment is the set a and the output response from the environment may be binary $(0, 1)$ [P-model], a finite collection of distinct symbols [Q-model] or a continuous output in the interval $(0, 1)$ [S-model]. In the present work we confine our attention to the P-model represented by the binary response of either a reward (0) or penalty (1). If the penalty probabilities from the environment do not depend on stage number n, the environment is classified as stationary; otherwise the environment is nonstationary. The c_i factors are of course assumed initially unknown. Recently, new mathematical models have been developed concerned with nonstationary environments in which the penalty probabilities associated with specific actions are functions of the probabilities with which the actions are selected.[95, 96] Such models are of practical significance since they correspond to the physical situation in which the availability of some resource is inversely proportional to the frequency with which it is utilised.

The convergence characteristics of learning automata are dependent on the properties of the algorithm used in the updating scheme. One quantity which is useful in determining the behaviour of the learning automaton is the average penalty received. Suppose at a certain stage n, if action a_i is selected with probability $p_i(n)$ then the average penalty conditioned on the total state probability vector $p(n)$ is given by

$$M(n) = \text{Exp}\{x(n)|p(n)\}$$

$$= \sum_{i=1}^{r} p_i(n)c_i$$

If the automaton actions were selected with equal probability then the average received penalty would be the arithmetic mean of the individual c_i's. Thus we define

$$M_0 = \frac{c_1 + c_2 + \ldots + c_r}{r}$$

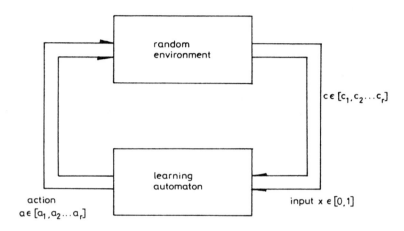

Fig. 4.1. *Interaction of learning automaton and environment*

A learning automaton is considered to learn if the average penalty is less than M_0, at least asymptotically. If the average penalty is less than M_0 the automaton is considered to be *expedient*. When an automaton is expedient it only does better than one which chooses its actions in a purely random manner. Thus if

$$\lim_{n \to \infty} \text{Exp}\,[M(n)] < M_0$$

the automaton is classified as expedient. In contrast if

$$\lim_{n \to \infty} \text{Exp}\,[M(n)] = \text{Min}_{i}\,\{c_i\}$$

then the automaton is said to be *optimal*.[93] Optimally implies that asymptotically the action associated with the lowest penalty probability is selected with probability one. Although this property may appear to be desirable it should be noted that in practice with nonstationary environments a suboptimal performance is preferable. In such situations we wish to avoid conditions in which the automaton locks onto a specific action. Suboptimal schemes are described as ϵ-optimal in which the degree of suboptimality ϵ may be varied by suitable choice of parameters in the reinforcement schemes. Thus an ϵ-optimal learning automaton satisfies the condition

$$\lim_{n \to \infty} \text{Exp}\,[M(n)] = \min_{i} \{c_i\} + \epsilon$$

Finally, if the expected penalty decreases monotonically at every stage n as given by

$$\text{Exp}\,[M(n+1)|p(n)] < M(n)$$

the automaton is said to be absolutely expedient.

The Russian Tsetlin in a pioneering paper[97, 98] described several fixed structure learning automata operating with linear tactics in random environments. Subsequently, several attempts have been reported to design unconditionally optimal learning automata.[99-101] Essentially, all such schemes may be described in terms of ergodic finite state homogeneous Markov chains and thus they are reasonably amenable to mathematical analysis.

The operation of a Tsetlin automaton may be understood by reference to Fig. 4.2 which shows a two action automaton with a memory size of n, and has one action corresponding to internal states 1 to n and the other corresponding to internal states $(n+1)$ to $2n$. When the automaton takes action 1 the environment outputs a stochastic sequence of value c_1, while action 2 corresponds to a stochastic sequence of value c_2. If the automaton receives a penalty the automaton moves towards states n and $(n+1)$ while, in response to a reward, the automaton moves towards end states 1 or $2n$. Thus, with output action 1 the automaton performs a simple random walk between its internal states, with a reflecting barrier beyond state 1 and with output action 2 the automaton performs a simple random walk between its internal states, with a reflecting barrier beyond state $2n$. If an action has associated with it a c_i, the value of which is greater than half, the automaton will tend to move towards states associated with the alternative action while, if the value of c_i is less than half, the automaton will tend to move towards the end state associated with the action it is already taking.

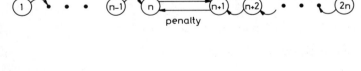

Fig. 4.2.   *Operation of Tsetlin automaton*

The operation of the Tsetlin automaton will fall into one of three modes depending on the environment. If the c_i's are about half, one action will tend to make the automaton move towards the corresponding end state. Thus one action is stable while the other is unstable and the automaton works well. If the c_i's are both greater than a half, both actions will tend to make the automaton move towards states associated with the other action. Thus both actions are unstable, the automaton moves between states n and $n + 1$ frequently and works poorly. If the c_i's are both less than a half, each action will tend to make the automaton move towards the end state associated with the action. Thus both actions are stable, with the automaton only moving from one action to another due to variance in the penalty probability causing it to be temporarily greater than a half over a long enough time to allow the automaton to move from one action to the other. If the larger penalty probability is not close to a half, or if the memory size is large, the automaton can output the wrong action for long periods of time and the automaton works poorly.

In order to overcome some of the disadvantages of the Tsetlin automaton, a simple modification was proposed by Krylov.[99] The Krylov automaton is very similar to the Tsetlin automaton in that it has a series of states 1 to $2n$, with states 1 to n being associated with one action and states $(n + 1)$ to $2n$ being associated with the other. It is in the movement between the states that the Krylov and Tsetlin automata differ as shown in Fig. 4.3. In response to a reward, the Krylov automaton acts as the Tsetlin and moves deterministically towards an end state but, in response to a penalty, the automaton acts in a stochastic manner and either moves towards states n and $(n + 1)$ or towards the end states with probability $\frac{1}{2}$.

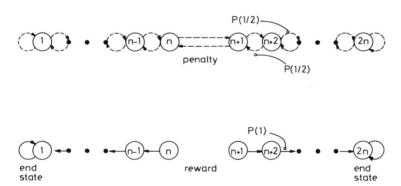

Fig. 4.3. *Operation of Krylov automaton*

The action of the Krylov automaton can be related to that of the Tsetlin automaton. If an automaton performs an action such that it receives a penalty with probability c_1 then

$$\text{penalty probability} = c_1$$

$$\text{reward probability} = 1 - c_1$$

If a reward response is taken as a movement towards states 1 or $2n$ and if a penalty response is taken as a movement towards states n and $(n + 1)$ then for the Krylov automaton

penalty response probability $= c_1$

reward response probability $= 1 - c_1$

and a similar argument applies to c_2.

Equating response probabilities we see that a Krylov automaton receiving penalty probabilities in the range $(0, 1)$ is equivalent to a Tsetlin automaton receiving penalty probabilities in the range $(0, \frac{1}{2})$. However, it has been shown above that the Tsetlin automaton does not function correctly with penalty probabilities both less than $\frac{1}{2}$ and as expected the Krylov automaton does not work well over the complete range of c_i's. To overcome these disadvantages recently attempts have been made to design automata which retain the desirable qualities of the Tsetlin automaton but also operate well for c_i's about any value rather than the value of half to which the Tsetlin automaton is limited.[102] The Krylov automaton took penalty probabilities which were greater than a half and produced penalty response probabilities which were less than a half. The modified automata to be described take two penalty probabilities of greater than a half but about a value c_m and, by using a stochastic response to a penalty, produce one penalty response probability which is less than a half and one which is greater than a half. Further, by using a stochastic response to a reward, two penalty probabilites both less than a half but about a value c_m will produce one penalty response probability which is greater than a half and one which is less than a half. This is illustrated in Fig. 4.4. Thus, provided c_m is known, any pair of penalty probabilities can be transformed to be about a half producing a Tsetlin type response.

The modified Tsetlin automata are similar to the Tsetlin automaton in that they have a series of states 1 to $2n$ with states 1 to n being associated with one action and states $(n + 1)$ to $2n$ being associated with the other. However, the movement between the states is more complex and is shown in Fig. 4.5. For the modified Tsetlin automaton, type 1 shown in Fig. 4.5a, and penalty probabilities about a c_m value greater than a half, as shown in Fig. 4.4a, to obtain penalty response probabilities c_1' and c_2' spaced about a half

$$c_m' = 0.5 \tag{4.1}$$

Using a stochastic response to a penalty with probability W_p of moving towards states n and $(n + 1)$ and assuming a deterministic response to a reward then

$$c_m' = c_m W_p \tag{4.2}$$

Substituting eq. 4.2 into eq. 4.1 gives

$$W_p = \frac{1}{2c_m}$$

W_p is to be a stochastic variable and so has a maximum value of 1 thus

$$W_p = \frac{1}{2c_m} \text{ if } \frac{1}{2c_m} < 1 \tag{4.3}$$

$$= 1 \text{ if } \frac{1}{2c_m} \geqslant 1$$

For penalty probabilities about a c_m value less than a half as shown in Fig. 4.4c, to obtain penalty response probabilities c_1' and c_2' spaced about a half again we must satisfy eq. 4.1. Using a stochastic response to a reward with probability W_r of moving towards the end state associated with the action output by the automaton and assuming a deterministic response to a penalty, an assumption justified by eq. 4.3 then

$$c_m' = c_m + (1 - W_r)(1 - c_m) \tag{4.4}$$

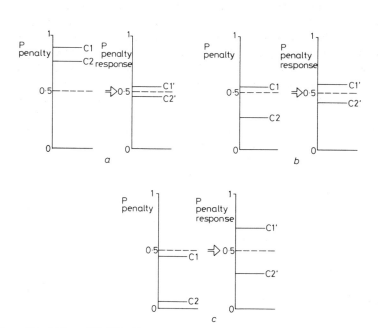

Fig. 4.4. *Effect of modified Tsetlin automata*

substituting eq. 4.4 into eq. 4.1 gives

$$W_r = \frac{1}{2(1 - c_m)} \text{ if } \frac{1}{2(1 - c_m)} < 1 \tag{4.5}$$

$$= 1 \text{ if } \frac{1}{2(1 - c_m)} \geqslant 1$$

For c_m greater than a half $W_r = 1$, so justifying the assumption made in forming eq. 4.2.

For the modified Tsetlin automaton, type 2 in Fig. 4.5*b*, in addition to penalty and reward responses we have an inaction response. If an inaction response is counted as half a penalty response, for penalty probabilities about a c_m value greater than a half as shown in Fig. 4.4*a* to obtain penalty response probabilities c'_1 and c'_2 spaced about a half we again must satisfy eq. 4.1. Using a stochastic response to a penalty with probability of W_p of moving towards states n and $(n+1)$ and $(1 - W_p)$ of remaining in the same state, and assuming a deterministic response to a reward then

$$c'_m = c_m W_p + \tfrac{1}{2} c_m (1 - W_p) \tag{4.6}$$

Substituting eq. 4.6 into eq. 4.1 gives

$$W_p = \frac{1 - c_m}{c_m} \text{ if } \frac{1 - c_m}{c_m} < 1 \tag{4.7}$$

$$= 1 \text{ if } \frac{1 - c_m}{c_m} \geqslant 1$$

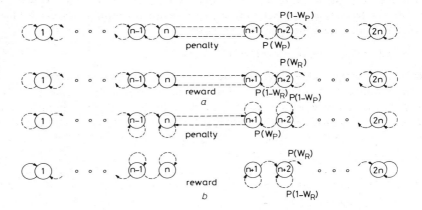

Fig. 4.5.   *Operation of modified Tsetlin automata*
    a   Type 1
    b   Type 2

For penalty probabilities about c_m values less than a half, as shown in Fig. 4.4*c*, to obtain penalty probabilities c_1 and c_2 spaced about a half again we require $c'_m = 0.5$. Using a stochastic response with probability W_r of moving towards the end-state associated with the action output by the automaton and assuming a deterministic response to a penalty, an assumption justified by eqn. 4.7, then

$$c'_m = c_m + \tfrac{1}{2}(1 - c_m)(1 - W_r) \tag{4.8}$$

Substituting eq. 4.8 into eq. 4.1

$$W_r = \frac{c_m}{1 - c_m} \quad \text{if} \quad \frac{c_m}{1 - c_m} < 1 \tag{4.9}$$

$$= 1 \quad \text{if} \quad \frac{c_m}{1 - c_m} \geqslant 1$$

For c_m greater than a half $W_r = 1$, so justifying the assumption made in forming eq. 4.6.

Eqs. 4.3, 4.5, 4.7 and 4.9 require a value for c_m. This is taken as the mean of estimated values for c_1 and c_2 obtained from two adaptive digital circuit elements (ADDIES) which respond to the reward/penalty signals obtained from the environment, these signals being fed to the ADDIE estimating c_1 when action 1 is output and to the ADDIE estimating c_2 when action 2 is output. It may be predicted that the type 2 automaton with the inaction response would have less variance than the type 1 automaton and would be more nearly optimal for the same memory size.

An alternative automaton structure which has been proposed is the so called modified estimating automaton (m.e.) The operation of the m.e. automaton consists basically of continually updating estimates of the penalty probabilities, where $g_i(t)$ is the estimate of c_i at time t. If the response $x(t + 1)$ is elicited from the environment by action a_i, where $x(t + 1) = 0$ denotes a reward and $x(t + 1) = 1$ denotes a penalty, then the estimate of g_i is updated

$$g_i(t + 1) = g_i(t) + w\{x(t + 1) - g_i(t)\} \tag{4.10}$$

where w is a constant whose inverse corresponds to the depth of memory.

The automaton then selects the lowest penalty probability estimate $g_i(t) = \min_j \{g_i(t)\}$ and action a_i is chosen with a finite probability $p(t)$. For simplicity only a two-state case, a_1 and a_2 will be considered. The probability of choosing action a_2 where action a_2 results in penalty probability c_2 being selected is

$$p_2(t) = \beta/2 + (1 - \beta) d(t) \tag{4.11}$$

$$p_1(t) = 1 - p_2(t)$$

$$d(t) = 1 \quad \text{if } g_2(t) < g_1(t)$$

$$= 0 \text{ otherwise}$$

where β is a small positive constant. Hence the automaton strongly favours action a_1 when it believes that penalty probability c_1 is smaller than c_2 and vice versa, i.e. action a_2 when $g_2(t) < g_1(t)$; action a_1 when $g_1(t) < g_2(t)$.

A survey paper[93] has considered the various variable structure schemes available. Variable structure stochastic automata have the potential of greater flexibility and may be analysed by means of Markov process theory.[104, 105] One basic scheme is the linear reward – penalty algorithm ($L_{r.p.}$) described for the simple two-state case by

reward (action a_1)

$$p_2(n + 1) = \alpha p_2(n) \qquad \text{where } 0 < \alpha < 1$$

$$p_1(n + 1) = 1 - \alpha p_2(n)$$

Penalty (action a_1)

$$p_1(n + 1) = \beta p_1(n) \qquad \text{where } 0 < \beta < 1$$
$$p_2(n + 1) = 1 - \beta p_1(n)$$

Thus, if the automaton selects an action a_i which results in a reward from the environment, the probability $p_i(n)$ is increased and $p_j(n)$ ($j \neq i$) is decreased. Similarly, a penalty response from the environment causes $p_i(n)$ to be decreased while $p_j(n)$ is increased.

An alternative scheme referred to as a linear reward – inaction algorithm ($L_{r.i.}$) is derived by simply ignoring penalty responses from the environment. Thus the action probabilities $p_i(n)$ remain constant during penalty responses. Other linear algorithms considered have included reward-reward, penalty-penalty and inaction-penalty schemes. In general, these schemes are inferior to the $L_{r.p.}$ and $L_{r.i.}$ algorithms. Both nonlinear and hybrid (linear/nonlinear) schemes have been suggested; the objective of such schemes is to obtain improved rates of convergence.[106]

4.2 Synthesis of learning automata

4.2.1 Fixed structure automata

The hardware design of Tsetlin and Krylov automata is very simple as shown in Fig. 4.6. The most significant bit of an up/down counter represents the action of the automaton and acts as input to the environment. The output of the environment and the action of the automaton are fed into combinational logic to convert these into an up/down control signal for the counter. The up/down signal is in turn fed into more combinational logic along with the state of the automaton and signals representing the memory size to provide a disable signal to prevent the counter exceeding the required memory size.

Fig. 4.7 shows the operation of a Tsetlin automaton with the central trace in each case indicating the switching instants for a reversal of penalty probabilities c_i. Fig. 4.7a shows the satisfactory operation of the automaton with c_i's of 15/16 and 1/16. Fig. 4.7b shows the effects of change of c_i's to 15/16 and 3/4, i.e. both greater than 1/2. It is evident that the automaton fails to operate. Finally, Fig. 4.7c illustrates the characteristics with c_i's of 3/16 and 1/16. In this case since the c_i's are both less than 1/2 the automaton again operates poorly and locks onto one action. These results are entirely consistent with the theoretical predictions.

Fig. 4.8a shows a Krylov automaton initially with output action 1, operating in a switched environment with c_i's of 0 and 15/16. As predicted the result is similar to a Tsetlin automaton working with both c_i's less than a half with the automaton locked into the output of one action. This locking is in fact a function of the memory size. The automaton has two stable states, with the state corresponding to the lower c_i being more stable than the other with stability increasing as the memory size increases. Variance in the penalty probabilities causes movement between the states and the time spent in a state depends on its stability. Thus while both states are stable, for small memory sizes, variance should cause movement between the states with the automaton spending more time in the most stable state. This can be seen in Fig. 4.8*b* which shows a Krylov automaton with memory size of 8 with c_1 of 7/8 and c_2 of 5/8 moving from states corresponding to c_2 to states corresponding to c_1, remaining in those states for a time then moving back. Finally, Fig 4.8*c* shows a Krylov automaton with memory size of 8 working in a switched environment with c_i's of 3/4 and 5/8. Since when the switching trace is high the automaton trace should be low it can be seen that it works poorly.

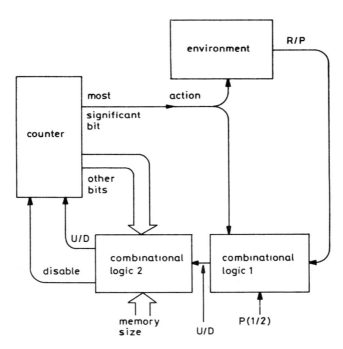

Fig. 4.6.   *Schematic diagram for Tsetlin and Krylov automata*

Fig. 4.9 shows results from a simulation of a modified Tsetlin automaton, type 1, operating in an environment with penalty probabilities of 0·6 and 0·9. It can be seen in Fig 4.9*a* that the automaton initially moves between actions 1 and 2 frequently but later moves to states associated with the action corresponding to the lower penalty probability. Initially, with the estimates of the c_i's in the ADDIES being zero, both actions are unstable but as the estimates of the penalty probabilities rise the actions become less unstable until in the steady-state the action corresponding to $c_{i(min)}$ is stable and the other unstable. The learning time is limited by the speed of response to the ADDIES. Between Fig. 4.9*a* and 4.9*b* the

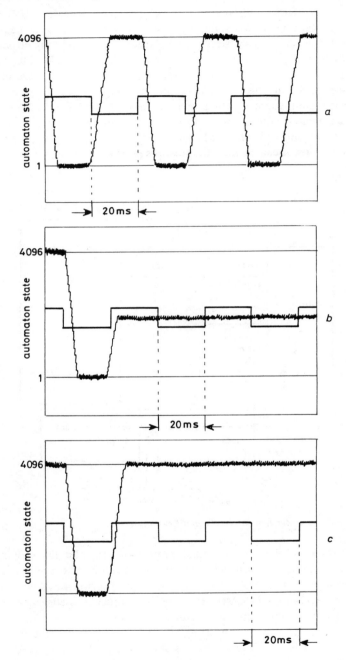

Fig. 4.7.   *Examples of operation of Tsetlin automaton in switched environments*

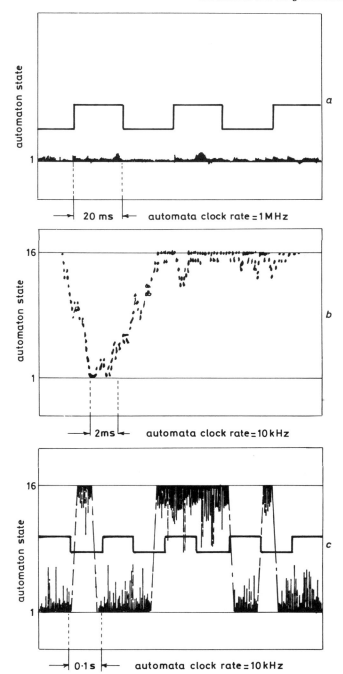

Fig. 4.8. *Examples of operation of Krylov automaton in switched environments*

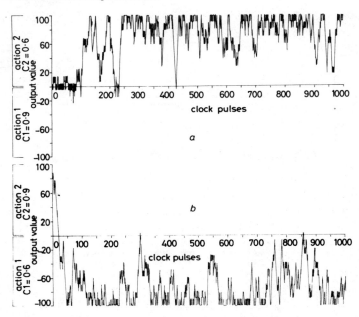

Fig. 4.9.   *Example of operation of type 1 modified Tsetlin automaton*

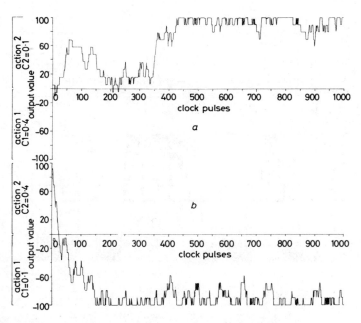

Fig. 4.10.   *Example of operation of type 2 modified Tsetlin automaton*

environment has been switched and it can be seen that the automaton reacts quickly to the change.[102]

Fig. 4.10 shows results from a simulation of a modified Tsetlin automaton, type 2, with the same parameters as the type 1 considered above but operating with penalty probabilities of 0·1 and 0·4, again with the environment switching between Figs. 4.10*a* and 4.10*b*. The automaton operates satisfactorily and the lower variance of the type 2 automaton can be seen. Fig. 4.11 shows results of simulations of Tsetlin and Krylov automata. Fig. 4.11*a* shows the Tsetlin automaton operating with penalty probabilities of 0·6 and 0·9 and moving frequently between states *n* and (*n* + 1), both actions being unstable, while Fig. 4.11*b* shows the Krylov automaton remaining in the incorrect state after a switch in the environment to 0·65 and 0·35.

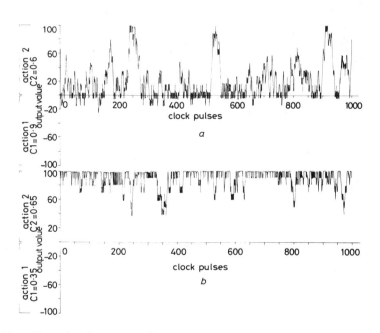

Fig. 4.11.   *Examples of operation of Tsetlin and Krylov automata*

The modified Tsetlin automata types 1 and 2 have been shown to be capable of good learning characteristics with no restrictions on the penalty probabilities that can be used, whilst retaining the short mean switching times and near optimal performance that characterises the Tsetlin automaton. It is hoped these automata will be of use in nonautonomous environments where their ability to reject actions that correspond to penalty probabilities above any value of c_m and their short switching times should prove valuable.

4.2.2 Modified estimating (m.e.) automaton

The schematic diagram of the m.e. automaton is shown in Fig. 4.12. Standard stochastic computing techniques are used for multiplication algorithms and the
generation of pseudorandom binary sequences (p.r.b.s.). The penalty probabilities
c_1 and c_2 are generated from a shift register maximal length p.r.b.s. generator and
stochastic adders and multipliers. The two c_i's are then routed to the appropriate
ADDIES by means of the steering circuitry.[107]

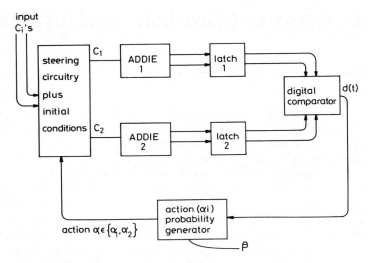

Fig. 4.12.   *Schematic diagram for m.e. automaton*

An estimate $g_i(t)$ of the penalty probability c_i may be simply obtained by means
of an ADDIE circuit. As described in Chapter 2 the basic exponential smoothing
equation may be written as (eq. 2.20)

$$S_t = S_{t-1} + (1 - \alpha)(A_t - S_{t-1}) \tag{4.12}$$

where the new estimate S_t is equal to the previous estimate S_{t-1} plus a correction
term multiplied by $(1 - \alpha)$. The value of α is governed by the number of states in
the up/down counter and for an N-state counter α is equal to $(1 - 1/N)$. Replacing
α by $(1 - 1/N)$ in eq. 4.12 we obtain:

$$S_t = S_{t-1} + \frac{1}{N}(A_t - S_{t-1}) \tag{4.13}$$

Comparing eq. 4.13 with eq. 4.10 we see that they are identical, with $w = 1/N$.

For a two-state system, c_1 and c_2 need to be estimated and this is accomplished
by ADDIES as described previously. In order to determine which of the two c_i's is
the smaller the two ADDIES contents are compared by a digital comparator whose
output $d(t) = 0$ if $c_1 \leqslant c_2$ and $d(t) = 1$ if $c_2 < c_1$. Prior to the comparator a latch

arrangement is used to enable the ADDIES to count to a steady-state condition and prevent any race-round conditions. In practice a 1 : 10 ratio of latch-sample clock frequency to main-system clock frequency was found to be adequate. Note that all clocks are synchronous. The output from the digital comparator $d(t)$ is used to generate a probability, $p_i(t)$ of choosing action α_i as dictated by eq. 4.11. The constant β may be used to alter the probability of choosing the correct action i.e. small β gives a correct action probability close to unity and vice versa. For experiments with nonstationary environments it is essential to have variable control over the factor.

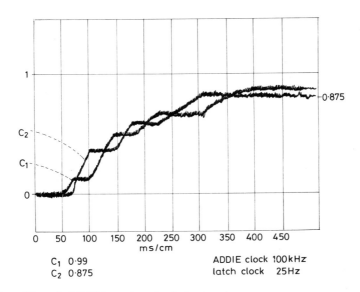

| C_1 | 0.99 | ADDIE clock | 100kHz |
| C_2 | 0.875 | latch clock | 25Hz |

Fig. 4.13.   *Alternate ADDIE count modes during learning*

The learning curves of Fig. 4.13 illustrate the alternate counting modes until a steady-state condition with $c_2 = 0.875$ is reached. In order to compare ADDIES of varying bit size the results of Fig. 4.14 were obtained. This shows the learning behaviour of a 12, 16 and 20-bit ADDIE system. As expected a compromise has to be made between speed of operation and accuracy in the penalty-probability estimate. If a very small bit size of ADDIE is used and the two c_i's are close together then there is the possibility that the wrong state will be chosen. This fact means that the automaton is no longer optimal, even with β equal to zero, thus contradicting the theoretical prediction of eq. 4.11. In practice a 12-bit ADDIE was found to provide a satisfactory compromise solution.

4.2.3 Variable structure automata

A two-state variable structure stochastic learning automaton (s.l.a.) can be implemented using a single ADDIE, as shown in Fig. 4.15. The contents of the ADDIE counter represent state probability $p_1(n)$, while $p_2(n)$ is simply the complement. An essential feature of the operation of the automaton is the updating of state probabilities in accordance with the environment or plant response. This is achieved in

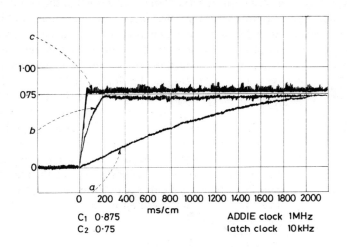

C₁ 0·875   ADDIE clock   1MHz
C₂ 0·75   latch clock   10 kHz

Fig. 4.14.   *Influence of ADDIE bit size on learning characteristic*

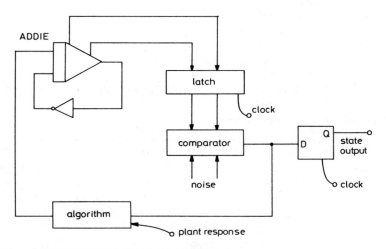

Fig. 4.15.   *Schematic diagram of two-state ADDIE s.l.a.*

the ADDIE s.l.a. by loading $p_1(n)$ from the ADDIE to a latch, and performing digital-to-stochastic conversion. The resulting stochastic pulse train is then transformed via the algorithm circuitry to an updated state probability $p_1(n + 1)$. The ADDIE then reaches an estimate of $p_1(n + 1)$, and after a suitable settling time, the next cycle can commence. A flip-flop on the comparator output represents the present state occupied, and state trajectories can be observed by filtering the output, or by direct digital analogue conversion of the ADDIE contents.[108]

The operating sequence for the ADDIE s.l.a. is as follows: The initial load operation sets up the requisite value of 0·5 (i.e. 'one – all zeros') in the counter, so that the output of the comparator is a stochastic sequence with an equal probability of 1's and 0's, representing random state selection at initial time t_0. At the first system clock pulse, this sequence is sampled and at the same time, the counter contents are copied into the latch. Then, when the clock pulse goes low, the punishment/reward signal resulting from the state of the D-type flip-flop is latched, and the ADDIE clock enabled, allowing the 'learning period' to commence. During this time, the ADDIE converges to the new value of $p_1(1)$, which is then used as the basis for the next cycle. The advantage of this design is that, since no locking-on problems can occur, it is possible to implement the more suitable ϵ-optimal schemes using methods of algorithm circuit design based on stochastic computing techniques.[109]

The ADDIE stochastic learning automaton enables several of the reinforcement schemes described previously to be implemented. For example, the linear reward/penalty scheme $L_{r.p.}$ may be expressed in two-state form as

(i) reward: (action a_1)

$$p_2(n + 1) = \alpha p_2(n)$$
$$p_1(n + 1) = 1 - \alpha p_2(n)$$

(ii) penalty: (action a_1)

$$p_1(n + 1) = \beta p_1(n)$$
$$p_2(n + 1) = 1 - \beta p_1(n)$$

where $0 < \alpha < 1$ and $0 < \beta < 1$.

Similar expressions hold for action a_2. Algorithm circuitry can be designed using a form of 'truth-table' as follows:

$p_1(n)$	$p_2(n)$	P/R	$p_1(n + 1)$
0	1	0	$\alpha p_1(n)$
0	1	1	$1 - \beta p_2(n)$
1	0	0	$1 - \alpha p_2(n)$
1	0	1	$\beta p_1(n)$

The design can obviously be extended to cover the ϵ-optimal linear reward-reward ($L_{r.r.}$) and reward-inaction ($L_{r.i.}$) schemes. The $L_{r.i.}$ scheme is particularly simple to accommodate, since the only modification required is to set the factor $\beta = 1$. The $L_{r.r.}$ scheme, in which the penalty is replaced by a lesser reward is given below in two-state form

(i) nonpenalty: (on action a_1)

$$p_1(n + 1) = 1 - \alpha p_2(n)$$
$$p_2(n + 1) = \alpha p_2(n)$$

(iii)   penalty: (on action a_1)

$$p_1(n + 1) = 1 - \beta p_2(n)$$
$$p_2(n + 1) = \beta p_2(n)$$

where $0 < \beta < \alpha < 1$.

As before, a truth-table is constructed to enable the algorithm to be translated into a circuit design:

$p_1(n)$	$p_2(n)$	P/R	$p_1(n + 1)$
0	1	0	$\alpha p_1(n)$
0	1	1	$\beta p_1(n)$
1	0	0	$1 - \alpha p_2(n)$
1	0	1	$1 - \beta p_2(n)$

By comparison with the truth-table for the $L_{r.p.}$ scheme, it is evident that the $L_{r.r.}$ circuit is realised simply by reversing the '$\beta p_1(n)$' and '$1 - \beta p_2(n)$' connections. The circuit arrangements for these linear schemes are summarised in Fig. 4.16.

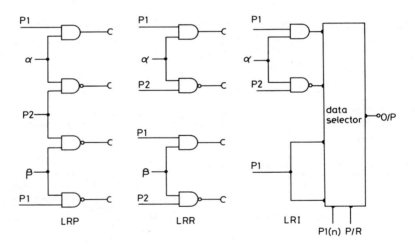

Fig. 4.16.   *Algorithm circuits for linear schemes*

Although the best of the linear schemes, the $L_{r.i.}$ scheme, has been widely reported as most suitable for many applications, considerable study has been made of nonlinear updating schemes. These tend to show faster initial convergence rates, and indeed one reason for the emphasis put on these schemes is to obtain optimum convergence times, especially when they are incorporated in hybrid schemes.[110] The simplest of the nonlinear schemes is that denoted as $N_{r.p.}^1$: which has 'square-law' nonlinearity.[93] It has been shown that this scheme is conditionally optimal, providing optimal convergence if $c_1 < \frac{1}{2} < c_2$ and expedient otherwise (c_i represents penalty probabilities). This scheme, again for the two-state case, is given below:

(i) nonpenalty (on action a_1)

$$p_1(n + 1) = p_1(n) + \alpha p_1(n)[1 - p_1(n)]$$
$$p_2(n + 1) = p_2(n) - \alpha p_1(n)[1 - p_1(n)]$$

(ii) penalty (on action a_1)

$$p_1(n + 1) = p_1(n) - \beta p_1(n)[1 - p_1(n)]$$
$$p_2(n + 1) = p_2(n) + \beta p_1(n)[1 - p_1(n)]$$

where $0 < \alpha, \beta < 1$.

The truth-table for the above scheme is as follows:

$p_1(n)$	$p_2(n)$	P/R	$p_1(n + 1)$
0	1	0	$(1 - \alpha p_2)p_1$
0	1	1	$1 - (1 - \beta p_1)p_2$
1	0	0	$1 - (1 - \alpha p_1)p_2$
1	0	1	$(1 - \beta p_2)p_1$

The circuit diagram for this scheme is shown in Fig. 4.17. The configuration is essentially similar to that of the $L_{r.p.}$ circuit, except that in the case of the $N_{r.p.}^1$ scheme, the constants α and β are replaced with the terms of the form $(1 - \alpha p_i)$ and $(1 - \beta p_i)$.

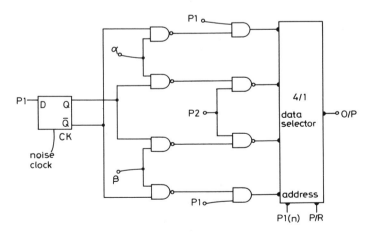

Fig. 4.17. *Algorithm circuit for nonlinear scheme*

Typical learning curves obtained with the $L_{r.p.}$ scheme using a main system cycle clock of 100 kHz are shown in Fig. 4.18. The curves clearly show how the degree of expediency increases as the reward-penalty 'δ' is increased from 1 to 64

$[\delta = (1 - \alpha/1 - \beta)]$, starting from an initial condition of random state selection at time t_0, i.e. $p_1(0) = p_2(0) = 0.5$. Learning curves obtained from the ϵ-optimal $L_{\text{r.i.}}$ scheme with $\alpha = 0.75$ are illustrated in Fig. 4.19. With this scheme, the automaton exhibits virtually full convergence to $p_1 = 1$ or $p_2 = 1$, according to the relative values of the penalty probabilities. The results presented here indicate learning times of less than 10 ms, pointing to the feasibility of practical, on-line operation.

4.2.4 Hierarchical-structure automata

For practical applications it is necessary to be able to design large state automata systems while preserving high operating speeds and simple circuit synthesis. One solution to this problem is to subdivide the state space and perform the random search between the states via a set of levels in a hierarchical structure. It has been suggested previously[111] that a multilevel approach can be used to overcome this problem of high dimensionality, and the application of simple two-level structures has been considered.[112]

The requirement for a memory capability in the automaton structure to establish a priority of state order during the learning period is embodied in the 128-state hierarchical system to be described.[113-114] A two-state s.l.a. 'cell' is time-shared between each location in a seven-level 'decision tree', and interfaced with a random-access memory (r.a.m.) to store intermediate probability values, as illustrated in Fig. 4.20. Since these seven two-state decisions are equivalent to one decision in a single-level 128-state automaton, the hierarchical structure give an enormous saving in hardware, and although it does involve more serial processing operations, there is not an excessive penalty in terms of operating speed. Another advantage of this configuration is that it can be made entirely modular in construction, which simplifies the design of very large systems. A further consideration is the implementation of the reinforcement scheme. With this design, it is possible to use the same two-state algorithm circuits as before, also time-shared between each decision level.

The two-state ADDIE s.l.a. which forms the 'cell' of the structure is essentially similar to that described earlier. The main difference is that now memory interface circuits are required, since the s.l.a. no longer acts in a continuous, selfcontained cycle, but operates instead in a time-shared mode with the 'decision tree'. The full operating cycle for this system is as follows: Initially, each memory location is set at 0.5, so that each decision has an equal probability of occurring; consequently, the probability of selecting any one state is $(0.5)^7$, i.e. 1/128. At each main sampling clock pulse, the cell output will be either 1 or 0, and this 'decision bit' is stored in a 7-bit 'state latch'. After a search through the decision tree, the state latch contents will therefore define uniquely one of 128 states, and also represent the output to the plant. The second half of the cycle consists of retracing the same path through the decision tree, this time applying a reinforcement scheme to alter the decision probabilities represented by the ADDIE in accordance with the plant response. Updating at each level is controlled by the punishment/reward signal (P/R) and the corresponding decision bit. The next full cycle can then commence, with revised decision probabilities and consequently a revised set of total state

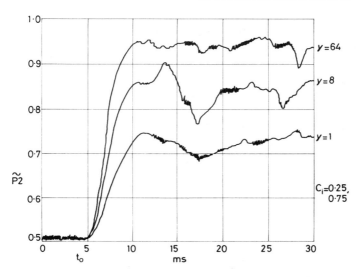

Fig. 4.18.   *Learning curves for $L_{r.p.}$ schemes*

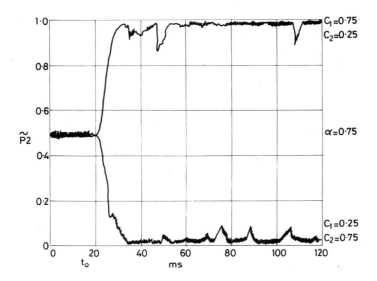

Fig. 4.19   *Learning curves for $L_{r.i.}$ schemes*

probabilities. As the learning process evolves, the decision path leading to the optimum state will be reinforced until, in the limit, assuming the use of an optimal scheme, all the decision probabilities along that path tend to unity.

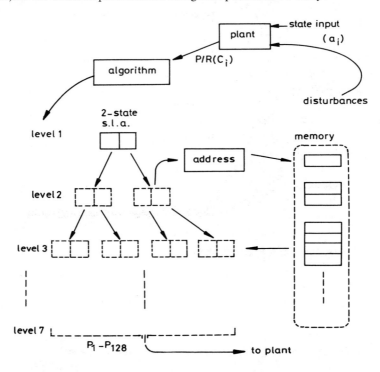

Fig. 4.20. *128-state hierarchical system*

The memory requirements are determined by the number of levels in the system. For each decision, the binary word in the ADDIE represents, together with its implicit complement, the decision probabilities for each 'direction'. The first decision level thus requires one word of storage, the second level requires two words, the third, four words, and so on. Therefore, the 128-state system requires a total r.a.m. allocation of $127 \times n$-bits ($n = 8 \rightarrow 12$). A central feature of the hierarchical system operation is the memory address procedure. Since each decision is a binary one, the design of the control circuitry is greatly simplified, and the address code is derived from the state latch, which is effectively a small 'scratch-pad' memory tracking the decision path.

The operating principle of the hierarchical system is illustrated in Fig. 4.21. This shows the learning behaviour of a three-level, 8-state system, with three simultaneous 'learning curves', obtained via a d.a. converter from the time-shared ADDIE, representing convergence to state 100_2 (i.e. state 4). The learning period is typically three times that expected of a single two-state system.

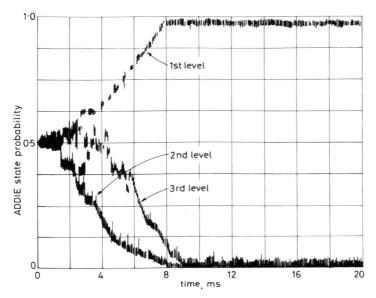

Fig. 4.21.   *Learning behaviour of 8-state system*

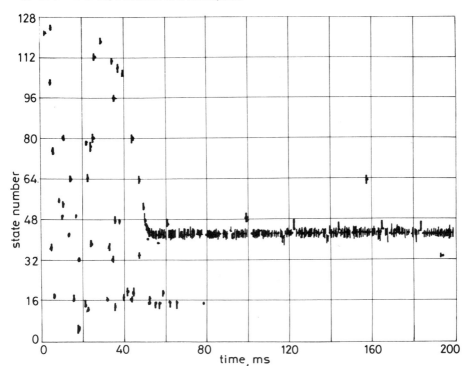

Fig. 4.22.   *Convergence of hierarchical system with steady state environment*

Static and dynamic optimisation experiments were carried out on the full 128-state system, using a simple simulated 'plant' in which one selected state carried a low penalty probability (0·25), and all other states a higher one (0·875). For these results, the system master clock was set at 2·5 MHz, a limit governed by the r.a.m. access time, and the system state was sampled every millisecond, with d.a. conversion of the state latch contents providing a 'map' of the state output. In Fig. 4.23, the optimum state is switched periodically from state 58 to state 106, and the automaton output is seen to follow the switching waveform within 50 ms. The transition interval is characterised by random selection between 'old' and 'new' states, until the new state finally predominates.

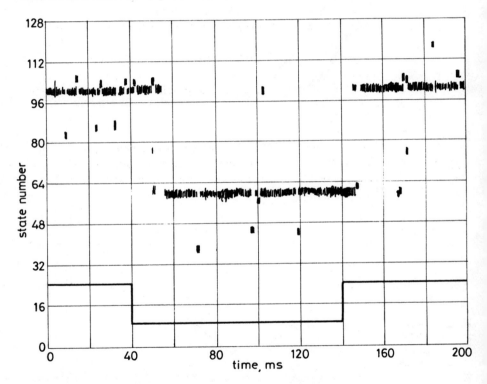

Fig. 4.23.   *Hierarchical system with switched environment*

4.3   Applications of learning automata

In spite of the significant advances made in learning automata theory, few applications to real problems have been reported. The possibility of using learning schemes in the optimisation of multimodal systems,[115-117] reliability,[118] power regulation in radio stations,[119] queueing systems,[120] detection and feature extraction in pattern recognition[121, 122] and adaptive routing in communication networks has been mentioned in the literature.[123–126]

Recently, the application of learning automata to problems involving task allocation in multiprocessor computer systems has been considered.[127, 128] Linear reward/penalty automata have been used as controllers for a three-node, three-processor system and their performance in terms of mean turnaround time compared to fixed scheduling disciplines. It has been shown that, for nonstationary system conditions, the learning automata scheme is relatively insensitive to environmental changes and provides significantly better performance than deterministic scheduling disciplines. These results are particularly important in adverse conditions.

In this section we will describe two applications of variable structure automata which are of particular significance:

(1) The use of learning automata for multimodal performance function optimisation in the presence of noise.

(2) The application of stochastic automata to the problem of adaptive routing of traffic in communication networks.

4.3.1 Global optimisation

A learning automaton is ideally suited to the problem of parameter optimisation of a noisy multimodal system (Fig. 4.24), since the inherent principle of random search avoids the affect of locking-on to local optima unavoidable with normal

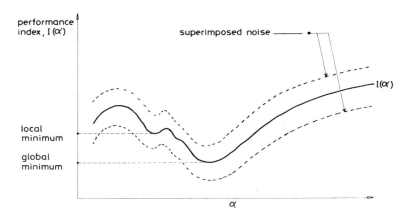

Fig. 4.24.   *Multimodal performance index with superimposed noise*

gradient methods. The automaton, in conjunction with a suitable interface, interacts with its environment in a manner analogous to a conventional feedback control system to evolve a 'suitable' final structure, as shown in Fig. 4.25.

In order to simulate multimodal environments it is convenient to use programmable read-only memories (p.r.o.m.'s) arranged to store performance index functions which have clearly defined global optima, local optima and saddle-points.[129] One such performance surface is illustrated in Fig. 4.26. The configuration of the simulated plant is shown in Fig. 4.27. The p.r.o.m. stores the c_i values as 8-bit numbers, each addressed by the appropriate action output from the stochastic learning automaton. The presence of noise on the surface is simply effected by interposing a full adder fed with noise derived from the central p.r.b.s. source. The

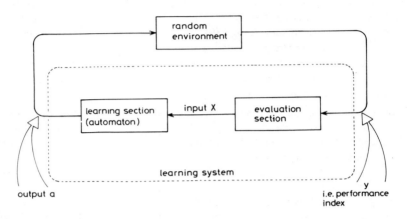

Fig. 4.25.   *Interaction of s.l.a. and environment*

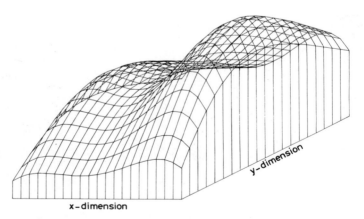

Fig. 4.26.   *Three-dimensional performance index surface*

resultant noise-corrupted value is then passed to a standard noise-comparator arrangement which produces a stochastic pulse train whose 'value' represents the current penalty probability. This is then sampled by the penalty/reward flip-flop to produce the appropriate system response (0:reward 1:penalty) to be fed to the algorithm circuit.

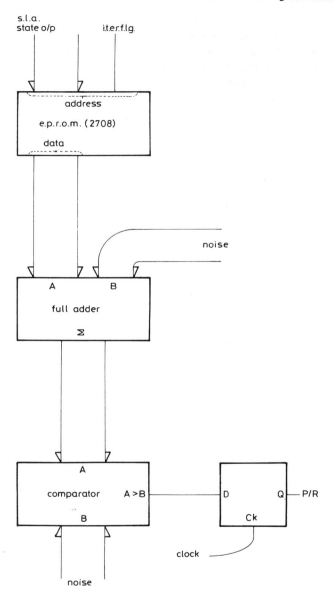

Fig. 4.27.   *Plant simulator with p.r.o.m.*

Extensive experiments have been performed with the hierarchical s.l.a. using the performance index described above with a superimposed noise component of $\pm\delta$ distributed uniformly over the surface.[148] Four bits of noise were in fact added, so that δ represented 8 units or approximately 3% of the full-scale range (0–255) of

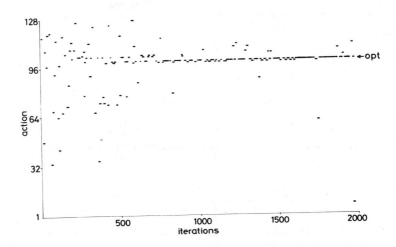

Fig. 4.28.   *Output map*

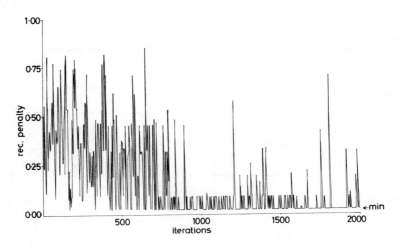

Fig. 4.29.   *Penalty curve*

the 8-bit c_i values. The output map, penalty curve and distribution curves for a typical simulation using the 128-state hierarchical automaton and an $L_{r.p.}$ scheme ($\alpha = 0.437$, $\beta = 0.992$) are shown in Figs. 4.28, 4.29 and 4.30, respectively. Note that convergence is obtained in around 1000 iterations. This particular learning run

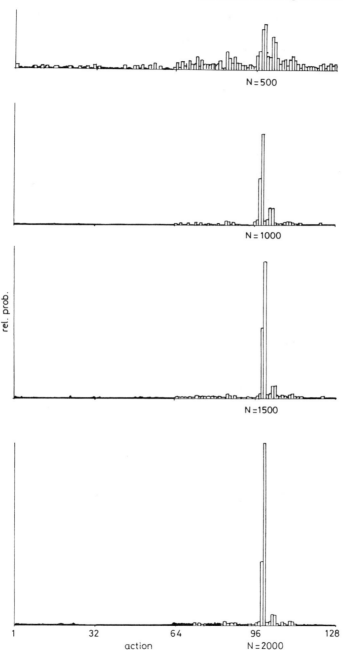

Fig. 4.30. *Distribution curves*

gives convergence to the optimum action (100), while the effect of spurious switching to suboptimal actions, shows up clearly on the penalty curve as transient 'spikes' to a higher level of received penalty.

The results of simulations clearly indicate the power of the hierarchical s.l.a. as a means of achieving rapid optimisation of a multimodal system, irrespective of contour, or of the presence of noise in the system. Even in cases where nonoptimal convergence occurred, owing to the use of a reinforcement algorithm where such behaviour is a known hazard, the automaton chose actions adjacent to the optimum and, therefore performed its allotted task of reducing the average received penalty, thereby achieving a corresponding improvement in system performance approaching the optimum value. It must be stressed that at no time did convergence to the local optimum occur, demonstrating that the s.l.a. has purely altitude sensitivity over the performance index as opposed to the gradient sensitivity of conventional hill-climbing methods. Results obtained with switched environments are particularly significant, since it is likely that many s.l.a. applications will involve nonstationary plant. The presence of noise on the performance index surface does not seem to impair significantly the performance of the s.l.a., and indeed it can be argued that its perturbating effect on the value of received penalty would help to dislodge a highly expedient learning scheme from an incorrect action to which it might otherwise lock-on. This would permit a slightly higher degree of expediency, which does have desirable features, to be catered for in the reinforcement scheme.

The development of the hardware hierarchical automata should permit the effective real-time control of several industrial plant systems which exhibit noisy multimodal performance characteristics.[130]

4.3.2 Adaptive routing in communication networks

Significant results for general communication network learning automata routing were first obtained in the field of military networks.[123, 124] Subsequent work considered the use of learning automata for adaptive routing in telephone networks.[125, 126, 131] Extensive simulations have shown that provided spare capacity is available in the network, then under conditions of overload and failure, learning automata routing techniques are superior to the fixed rule routing used currently in telephone networks.[132–134] This is of great importance since, owing to the very high investment in communications facilities, slight improvements in performance are economically significant.

In order to explain the operation of learning automata in a telephone network, we first deal with the fixed rule alternate routing method currently used in practical networks. Consider the simple five-node network of Fig. 4.31. Calls with source node A and destination node E would first chose the direct route AE. If this route is busy, alternative routes via nodes B, C and D may be selected. One problem which can occur is the classical 'ring-around the rosie' condition in which calls oscillate between nodes. The problem may be avoided either by encoding information which indicates the route taken by a call or by restricting the choice of permissible routes at each switching centre. For example, permissible routes from A to E in Fig. 4.31 would be in the order of $AE, ADE, ACE, ACDE, ABE, ABDE, ABCE$ and $ABCDE$.

Long distance telecommunication is performed on networks connected in a hierarchical array as shown in Fig. 4.32.[135] In the hierarchy, each primary switching centre P is associated with a sectional centre S which in turn connects to a higher level regional centre R. The alternate routing in the network is arranged to avoid 'ring-around the rosie' problems. For example, calls from $P1$ with destination $P2$ would first attempt the direct route $P1 - P2$. If this route is blocked, the route $P1 - S2 - P2$ would be attempted. Should this route also be blocked then a final choice at $P1$ would be the route $P1 - S1 - P2$. Similarly, as calls propagate up the hierarchy the various switching centres have a finite ordered choice of routes available.

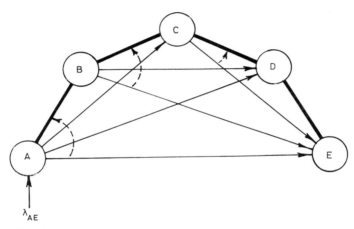

Fig. 4.31.   *Simple 5-node network*

Learning automata routing[131] operates by replacing the alternate routing schemes at each switching centre by automata. Two schemes are possible. If the source of origin of the calls is known at node k an automaton A_{ijk}, routes calls from node i to node j by determining the order in which the r allowable paths from node k are to be selected. For the more practical case in which the call source information is not known, the automaton A_{jk} at node k processes all calls with destination j. Each action of the automaton can correspond to either the selection of a specific trunk group or to a particular sequence of selections of trunk groups. Obviously, for r allowable trunk groups, $r!$ sequences are possible, and this can represent a fundamental restriction on the viable size of automaton to be used. As an example consider the hierarchical network of Fig. 4.32. Three alternative paths exist from switching centre $P1$. Thus a six action automaton would be required at $P1$ with each action corresponding to a particular ordered sequence of the three paths. It should be noted that it would be normal practice always to choose the direct route first before applying the learning automata routing scheme. Thus, for centre $P1$, if the direct route to $P2$ was blocked a two action automaton would be used to select the two possible sequence of routes from $P1$.

Fig. 4.33 shows a 12-node hierarchical network simulated by a recently developed

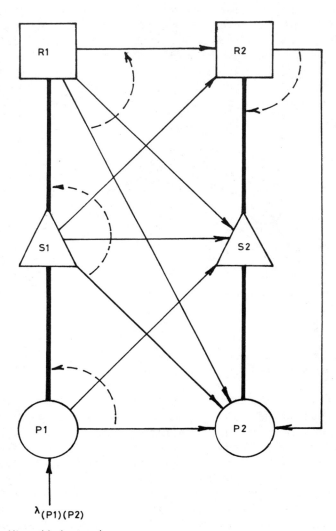

$\lambda_{(P1)(P2)}$

Fig. 4.32.   *Hierarchical network*

microprocessor based simulator.[134] Typical curves of blocking probability as a function of number of calls for both fixed rule and automata routing using $L_{r.i.}$ automata are shown in Fig. 4.34. The useful figure of merit, blocking probability, is simply the ratio of the number of calls blocked to the number attempted. For 2000 calls

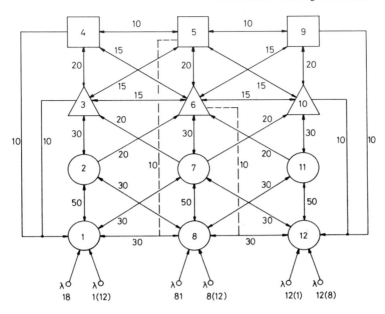

Fig. 4.33.   *12-node hierarchical network*

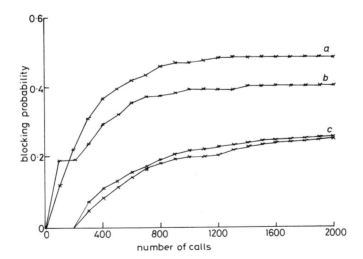

Fig. 4.34.   *Simulation results for 12-node network*
   a Fixed rule ($I_{78} = 0$)
   b Learning automata ($I_{78} = 0$)
   c Learning automata and fixed rule (normal conditions)
   $\lambda_{18} = \lambda_{81} = \lambda_{8(12)} = \lambda_{12(1)} = 10$ calls/min

the fixed rule gave a total blocking probability of 0·25 with 500 calls blocked at nodes 7 and 8. In contrast, the automata scheme gave a total blocking probability of 0·21 with 420 calls blocked at nodes 2, 7, 8 and 11. It should be observed that although both schemes gave similar total blocking probabilities the blocking probabilities for the individual sources are found to be more evenly distributed in the case of the automata scheme. This equality of service is an important design consideration in telephone networks and it is a characteristic of learning automata that they attempt to provide an even distribution. Also shown in Fig. 4.34 is the behaviour of the network under a condition of link failure between nodes 7 and 8. For this simulation, the fixed rule gave a total blocking probability of 0·496. Out of 2000 calls 992 were blocked at nodes 7 and 8. The automata scheme gave a total blocking probability of 0·393 with 785 calls also blocked at nodes 7 and 8. As expected the significant reductions in individual blocking probabilities occurs for sources λ_{18} and $\lambda_{12(8)}$.

It is likely that future developments in telecommunications switching associated with stored-program computer techniques will cause a gradual movement away from the hierarchical-network structures to more general topological networks. It is thus of considerable interest to investigate the performance of automata routing techniques in general networks. As a specific example we will consider the network shown in Fig. 4.35 in which calls arrive at node 1 destined for node 7. Typical simulation results are shown in Fig. 4.36. Under normal conditions the fixed rule is clearly suboptimal since all the calls which are blocked (596) are blocked at node 3. A mixed rule is needed at node 1 to utilise the available network capacity. For the fixed rule the effective number of trunks in the network is 45 and the minimum blocking probability that can be achieved is 0·296.[132] The actual experimental value for blocking probability was 0·298. For the $L_{r.i.}$ automata scheme the available trunk capacity is 60 corresponding to a minimum network blocking probability of 0·113. This compares very favourably with the experimental value of 0·131 corresponding to 262 blocked calls at nodes 2, 3 and 8. In the case of the link failure between nodes 6 and 7 using the fixed rule all blocked calls (1360) are blocked at node 6. The effective trunk capacity for the fixed rule under this failure condition is reduced to 20 trunks with a corresponding minimum blocking probability of 0·67. Again this compares very favourably with the experimental value for blocking probability of 0·68. In the case of the automata routing the effective number of trunks is 55 and the total blocking probability is reduced to 0·269 with 538 of the 2000 calls blocked at nodes 2, 6 and 8.

The following general conclusions may be made on the application of learning automata techniques to hierarchical and general network structures:

(1) Under normal operating conditions the $L_{r.i.}$ automata scheme performs at least as well as the optimum fixed rule.
(2) The automata routing acts in such a manner as to provide equality of service for the various individual call sources.
(3) Provided that additional capacity is available within the network, for abnormal operating conditions including link failure, node breakdown and focussed overloads,

the automata scheme has been shown to provide significantly reduced blocking probabilities and node congestion.

Packet-switched networks are distinguished from circuit-switched networks by the fact that the information to be transmitted is decomposed into finite packets

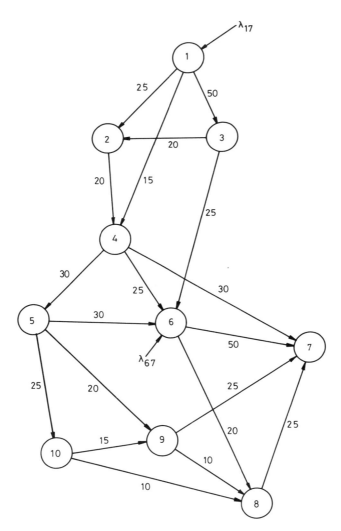

Fig. 4.35.   *10-node general-mesh network*

which are transmitted from node to node until the destination is reached. A large number of techniques have been proposed for routing in packet-switched networks. The strategies can be broadly classified into centralised, distributed and hybrid techniques. Distributed methods include flooding,[136] random routing,[137-138] 'short-est-queue bias'[139] and backwards learning.[136] Examples of centralised algorithms are multicommodity flow,[140-142] fixed multiple-route assignments,[143, 144] proportional

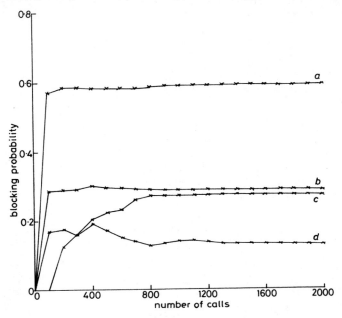

Fig. 4.36.   *Simulation results for 10-node network*
    a Fixed rule (fault condition)
    b Learning automata (fault condition)
    c Fixed rule (normal conditions)
    d Learning automata (normal conditions)
    $\lambda_{18} = 10$ calls/min, $\lambda_{67} = 0$

routing[145] and adaptive routing.[146] An example of hybrid routing is the so-called 'delta routing'[147] in which a combination of global and local decisions is implemented in such a manner as to achieve a consonant global strategy, while individual nodes are free to respond to instantaneous changes in their local environments. The possible application of learning automata routing to packet-switched networks appears to be a very promising approach. Essentially, learning automata routing at each node could be viewed as a combination of random and proportional routing with the added local learning ability providing very fast reactions to changes in the immediate environment. Such schemes as with the telephone network should provide fast adaptive response to changes in operating conditions including traffic overloads and network facility failures.

Chapter 5

Theory and applications of burst processing

To multiply two numbers with 7-bit accuracy, a 2000 transistor microprocessor needs about 10 clock-periods. A stochastic computer gives the same accuracy (within one standard deviation) in 10^4 clock periods — but it only uses a two transistor AND for the operation. *Burst processing* is a compromise solution which uses deterministic methods (like a microprocessor), but averages (like a stochastic machine).[149] We demonstrated in Chapter 1 that the fundamental idea is to perform 1-decimal digit arithmetic and to obtain accuracy by averaging over various inputs. For example, the number 3·4 would be treated as 4444333333 and 6·2 as 7766666666. Using the insight that the average of a sum is the sum of the averages, we can use a simple adder which forms (4 + 7), etc., the average being $9·6 = 3·4 + 6·2$.

Before we consider the fundamental principles of burst processing, it should be clearly stated that this method is not a panacea. As a matter of fact it is somewhat wasteful of bandwidth. For example, in the 10-slot version we could represent the integers 0 through 10 by 4 weighted binary digits. However, the advantages of burst processing may well make this sacrifice acceptable.

Among these advantages are:

1. Precision increases linearly with the number of time-slots; no correlation difficulties.
2. Simple computational circuitry (typically less than 10% of weighted binary systems).
3. Possibility of online processing without staticising the information; No synchronisation is necessary.
4. Significant error tolerance because of averaging.
5. Availability of a first approximation at all times.
6. Novel techniques for comparing, sorting, maximum-operations etc., leading to low-cost equipment.
7. Novel techniques for digital filter synthesis.
8. R.C.-element emulation without capacitors, an important point for integrated versions.

It should be noted that burst processing is only efficient if we can use crude approximations for our averaging process. Happily enough such crude approximations are perfectly accpetable in several important application areas including transmission of audio and video signals, in control-system computers (in which the output is averaged), in ranging and windowing equipment (sonar, radar) and in sampling devices for pure digital radios.

This Chapter describes purely digital techniques for performing arithmetic using bursts and the use of noncompacted bursts, and the corresponding encoders, is considered. The general area of digital filters, realised in burst technology, is discussed and practical examples from the area of convolution, adaptive filtering and r.f. tuning and demodulation are given. Finally, applications to picture processing and 'spatial' versions of burst processing are presented.

5.1 Encoding techniques

The basic ideas and some simple circuit elements for burst processing were introduced in Chapter 1. A possible format for representing the integers from 0 to 10 is a 10-slot frame (called a 'block') with the integer n corresponding to n adjacent pulses at the beginning of the frame. This will be called a 'compacted burst'. Instead of using 10 slots one can, of course, also use m-slots per block; systems have been built with $4 \leqslant m \leqslant 16$. If we do not mention the value of m explicitly, it will be assumed that $m = 10$. One of the interesting properties of this representation is that a 'window' of length m always contains the same number of pulses, (namely n) as long as we continue to represent 'n', quite independently of whether the beginning of the window coincides with the beginning of a block or not.[11]

The so-called block sum register (b.s.r.) considered in Chapter 1 (Fig. 1.17) exploits the above 'windowing property' in that the (quantised analogue or m-state, here 10-state) output is proportional to the number of 'ones' in the shift register. Note that by adjusting V we can actually multiply this number of 'ones' by a predetermined coefficient; this possibility is exploited in building digital filters using b.s.r.s.

It must be mentioned that the compacted burst format, although usually easier to handle, is by no means a necessity. *Uncompacted bursts* have 'ones' scattered throughout the block and thus lose the sense of a 'burst' of ones. Better frequency response is attainable with uncompacted bursts, but some of the cheap error correcting methods possible with compacted bursts are lost.[150] Fig. 5.1 shows a variety of possible encoders.[151-153] The first one (Fig. 5.1a) is the original ramp encoder which consists of a b.s.r. connected as a stairstep (or staircase) generator and a comparator. It does actually furnish compacted bursts. Because the stairstep generated by the b.s.r. samples the input at a time dependent on the input voltage, the input must change by less than one level of quantisation during one stairstep. This obviously limits the maximum input frequency. Specifically, a 1 V peak-peak sine wave of frequency f_c with maximum slope around $t = 0$ must vary less than one level of quantisation $1/N$ (for N-bit bursts and a 1 volt stairstep). Therefore, for bit rate $f_s(t = \pm 1/2f_s)$,

$$1/N > 0{\cdot}5\left(\sin\frac{N2\pi f_c}{2f_s} - \sin\frac{-N2\pi f_c}{2f_s}\right)$$

or where F_s is the normalised sampling frequency f_s/f_c

$$F_s > N\pi/\text{ARCSIN}(1/N)$$

For large values of $N \times F_s$ we have

$$F_s > N^2\pi$$

With the addition of sample/hold circuitry as shown in Fig. 5.1*b* the bit rate can be reduced to the level of the Nyquist sampling criterion, i.e. samples at twice the signal frequency. Hence

$$F_s > 2N$$

A different encoder, using essentially the same components, is the delta block encoder (Fig. 5.1*c*). Here the b.s.r. tracks the input signal by clocking in 'ones' or 'zeros' depending on whether it is below or above the input. Unfortunately, the b.s.r. only changes its output if the input bit is the opposite of the final bit and the tracking depends on the history of the input signal. Fig. 5.1*d* shows a possible solution. We use a bidirectional b.s.r. with a 'one'-feed on the left and a 'zero'-feed on the right. Fig. 5.1*e* shows a not uncommon 3-level version called 'bibursts'[153] which has synchronisation and bandwidth advantages. Here we transmit a positive or negative pulse when the input differs from the b.s.r. value by more than half a step, and, therefore, the receiver does not need a clock. In fact the clock is derived from the edge of the input pulses and simply freezes its value if no input pulse arrives. The average bit-rate requirement is obviously reduced for biburst systems. For a 1 V peak-peak signal with cutoff frequency f_c, the average absolute value slope is f_c. With step size $1/N$, the system need transmit only N-bits per cycle. A normal delta modulation system would transmit $F_s = f_s/f_c$ samples per cycle. Therefore a biburst system uses only $N/F_s \times 100$ percent of the bits of a standard delta modulation system under these conditions.

One of the advantages of burst processing is that error-pulses can be reasonably well absorbed because the b.s.r.s show averages. In case we use compacted bursts, one can improve the situation by taking a majority vote for every three adjacent slots. This procedure fills in 1-slot gaps and suppresses single pulses. It has been shown that under some noise conditons (10% error rate and above) this system gives results superior to Read-Muller codes.[150] In the case of compacted bursts we can transmit pulses of varying width constituting the envelope of the bursts, that is, burst processing is quantised pulse-width modulation. On the receiving end we then simply AND the envelope with the systems clock.

5.2 Arithmetic units

Representation of positive and negative numbers in burst processors has been accomplished in two ways. The sign and magnitude representation uses an additional

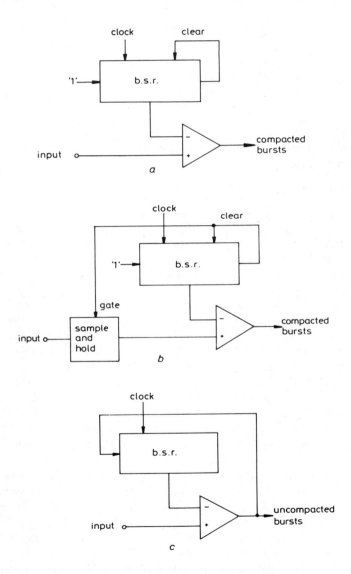

signal to represent the sign of the number, the magnitude is the same form as an un-signed burst. A *biased* burst represents signed numbers by assigning the value of zero to a half-filled burst; less than half is negative, more than half is positive.

Apart from the simple circuits described in Chapter 1,[164, 166, 167] logical arithmetic units of many different designs have been built. One of the more sophisticated versions uses the so-called carousel multiplier/divider (c.m.d.)[154, 155] shown in Fig. 5.2. The fundamental idea of the c.m.d. is that both multiplication and division

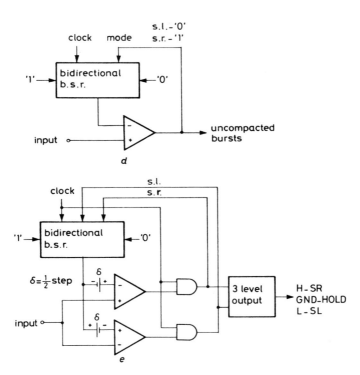

Fig. 5.1.   *Burst encoders*
    a   Ramp encoder
    b   Improved ramp encoder
    c   Delta block encoder
    d   Improved delta block encoder
    e   Biburst encoder

of burst sequences can be thought of as the 'filling up' of a register (called *Modular Register* in Fig. 5.2) by a stream of pulses, being careful to note the occurrence of overflow. In case of the multiplication of A and B, we make the modular register of length 10 and shift into it as many blocks of B as indicated by A: counting the number of overflows, we then obtain $AB/10$. Note that if we leave the modular register uncleared at the end, we shall obtain an exact value when we average.

For division A/B we simply make the modular register of length B and note its overflow as we shift in a fixed number (e.g. ten) blocks of A. Of course, only the 'ones' in A produce the filling of the modular register. The slightly exotic design of

Fig. 5.2 actually adjusts the length of the modular register dynamically by walking the 'ones' in B through a *Selector Register*. That portion of the modular register opposite the ones in the selector register is the active part of the modular register. By rotating 'load' and 'unload' gates (determined by detecting the 01 and 10 edges of the rotating active part) we transfer the 'ones' of A into the rotating active part

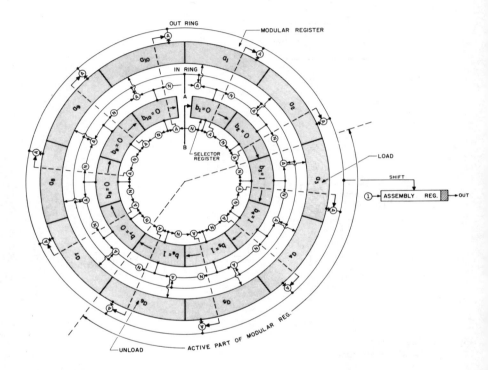

Fig. 5.2.   *Carousel multiplier/divider*

of the modular register and also detect the overflow. The situation can be visualised by thinking of the flip-flops in the active part as freightcars on a circular track, the train being filled up — while it moves — from the last car forward and emptied, as a block, whenever the car next to the locomotive is filled. The reason that the dynamic adjustment of the modular register is desirable is that we then actually form $(\overline{A/B})$. Note that often we required $(\overline{A/B})$. If we impose the condition $A/B \geqslant 0.5$, it can be proved that the two averages differ by less than 5% in the 10-slot system.

It has been shown previously that it is possible to modify a serial full-adder in such a fashion that it furnishes an uncompacted sum of two (uncompacted) bursts with a normalising factor of $1/2.$[150, 155] Fig. 5.3 shows why we call the arrangement a *perverted adder* (p.a.). What is normally the carry output of a serial adder gives the (normalised) sum while the sum output in normal operation becomes the input

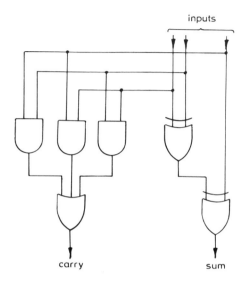

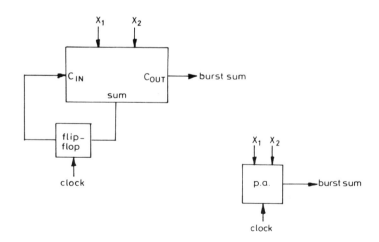

Fig. 5.3. *Perverted adder (p.a.)*

to the carry-delay circuit. The carry-delay is in effect a master/slave D-type flip-flop which delays the carry by one clock cycle and then reinserts it in the next cycle of the adder. Although the carry feedback limits the speed, there are also very attractive features, in particular the fact that we can cascade many p.a.s if necessary. The p.a. circuit is all that is required to add two unsigned bursts or two bursts with biased burst representation. To subtract biased bursts, an inverter is inserted in the input lead of the subtrahend. The output is always a properly scaled, uncompacted burst, and uncompacted bursts are completely acceptable as input.

The ease with which p.a.s can be cascaded is used in the p.a.-multiplier of Fig. 5.4. Here we use a tree of p.a.s to (dynamically) add all digits in A whenever B has a 'one'. The figure shows the arrangements for an 8-slot burst format, because this is more efficient than the standard 10-slot format for a tree-structure, but it is, of

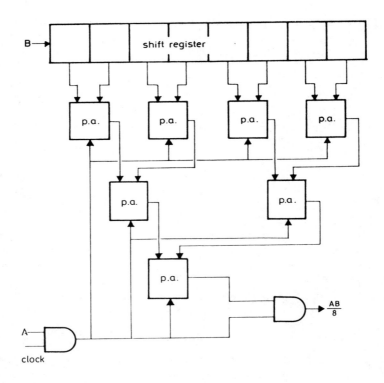

Fig. 5.4.   *P.A. multiplier*

course possible to design a 10-slot version. The multiplicand ('B' in Fig. 5.4) flows through the shift register at the top of the diagram. This shift register is exactly the length of one burst, and thus, like a b.s.r., always contains as many ones as the value of the number being represented. It should be observed that the multiplier ('A' in Fig. 5.4) is ANDed with a clock signal and the output of this AND gate is used to clock the perverted adders. The top row of p.a.s will produce a scaled sum of the contents of the shift register, and this sum will appear in parallel on the four outputs from the top row. Actual addition takes place only when a one is present on A to produce the clocking signal, with the result that B will be added to the p.a. tree A times and the output will be the scaled sum of these additions. Each level of the p.a. tree causes a scaling by a factor of 2, hence the factor of 8 for the multiplier in Fig. 5.4. Delay time through this circuit is dependent on the burst sum propagation. Recalling that this is actually the carry output as the full adder is normally used and that the output is often faster than the production of the sum, propagation speed through the tree is quite good.

A p.a. divider can be constructed by adding only a few components to the multiplier. Recall from the discussion of the p.a. multiplier that the p.a. tree produces one output pulse (on the average) for every 8 'ones' added into the top row of p.a.s. If one were to 'load' the tree with $8\text{-}n$ 'ones', then just n more 'ones' would produce a 'one' at the output. This would, in effect, be performing division by n on the input stream supplying the additional pulses. If every time the output of the tree produces a one, we load the tree again, we are performing a continuous division by n. The number of ones needed to load the tree ($8\text{-}n$) is easily generated by inverting the divisor and passing it into a shift register. A divider of this type needs a two phase clock. During one clock phase the inverted divisor is added into the tree (when needed), and during the other phase the dividend is added into one input of one p.a. in the top row. The p.a. tree must be clocked on both phases.

It is possible, at the expense of some hardware to build a divider using a single phase clock.[155] This divider runs at speeds comparable to the multiplier as it is necessary to wait for only one ripple through the tree each cycle instead of two. This can be performed in either of two ways. The inverted divisor can be staticised (i.e. held without shifting for 8 clock cycles) in a register in such a way that the 'one' bits in that register do not interfere with the input of the dividend. It may also be accomplished by moving the 'one' bits (specifically the one that falls on the lead reserved for the dividend) as necessary with combinational logic. This latter approach is illustrated in Fig. 5.5. The quotient in the divider of Fig. 5.5 is not scaled, and therefore overflow occurs only when we attempt to divide by numbers smaller than unity. Division of any number by zero yields bursts completely filled with ones.

Addition and subtraction of biased bursts involves trivial implementation. Multiplication and division of sign and magnitude bursts is also very simple. The only additional circuitry needed is an exclusive-OR gate to process the sign information. The most economical way to perform division on biased bursts appears to involve conversion to sign and magnitude and back again. In each of these circuits it is necessary to staticise bursts in registers, and the data is delayed for a total of only one or two burst periods.

A possible design for a sign and magnitude p.a. adder is shown in Fig. 5.6. The principle components are two p.a.s and an $n + 1$ bit bidirectional shift register. Not shown are an n-bit parallel-load shift register and the clocking circuitry to gate the result from one register to the other and to clear the bidirectional shift register after every burst. This design is in fact capable of accepting compacted or

uncompacted bursts and transmits compacted bursts.[155] The four inputs to this
adder (two bursts and two signs) are duplicated above each of the p.a.s in Fig. 5.6.
Logic gating between the input and the p.a.s causes the left p.a. to add only positive

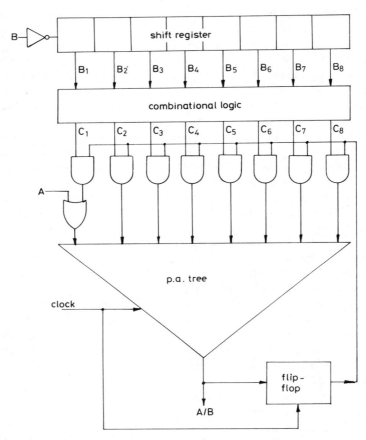

Fig. 5.5.   *P.A. divider*

numbers, the p.a. on the right adds only negative numbers. The outputs of the p.a.s
are used to control the shifting of the bidirectional shift register, and the shift
register is end connected via inverters. This ensures that the number of ones in the
register corresponds to the magnitude regardless of the arrival times of positive or
negative addend or augend. The left most bit of the register always contains the sign
bit according to the convention of a one for negative. The eight output bits are
gated out in parallel, one of these bits being generated by the OR of the two end
bits. It has been shown that by adding additional circuitry to the p.a. multiplier of
Fig. 5.4 it is possible to multiply biased bursts directly wth p.a.s.[155]

It should be noted that representation conversion can be simply accomplished. Fig. 5.7 shows a sign and magnitude to biased converter and a converter from biased to sign and magnitude representation is given in Fig. 5.8.

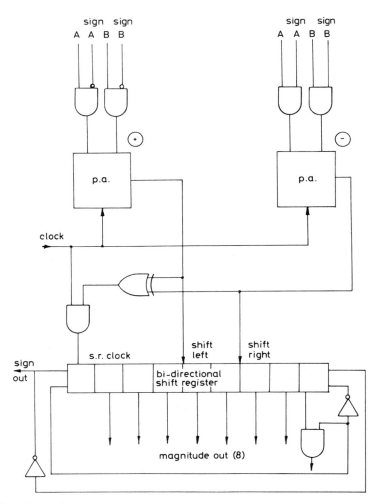

Fig. 5.6.   *Sign and magnitude adder*

We have not commented on the complexity of arithmetic units for burst processing. In fact by optimising the design it has been possible to imitate an 8-bit microprocessor (10 MHz clock) with about 10% of the gate count.[153, 155] In these times of ultracheap hardware this may not be so impressive, but it should be noted that a simplification by a factor 10 means also 10 times less power consumption, 10 times less space, and 10 times more reliability.

5.3 Applications

5.3.1 Burst digital filters, convolution, prediction theory and radios

Like stochastic processing, burst processing was initially a method in search of an application. For burst processing, one of the more promising fields is that of digital filtering. Fig. 5.9 shows why such an application is desirable. When averages of many (more or less random) numbers are calculated, each number can be a crude approximation, in particular a 8- or 10-slot burst approximation.

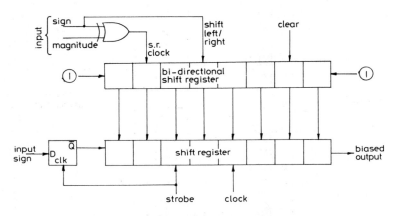

Fig. 5.7.   *Sign and magnitude to biased burst converter*

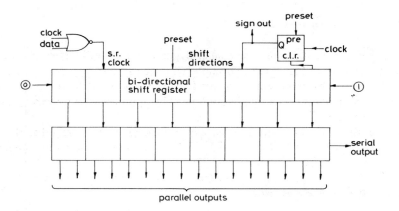

Fig. 5.8.   *Biased burst to sign and magnitude converter*

Recent work has concentrated mostly on realising f.i.r. (finite impulse response) filters, i.e. filters in which delayed versions of the signal are multiplied by certain constants and then added.[150] No feedback from output to input is allowed as in i.i.r. (infinite impulse response) filters. The reason for this limitation is that the noise problems in i.i.r. filters are much more severe and the (crude) burst approximations are therefore less attractive. Nonetheless, i.i.r. filters have been shown to be viable, for instance in the PREDICTORBURST project discussed below. The theory

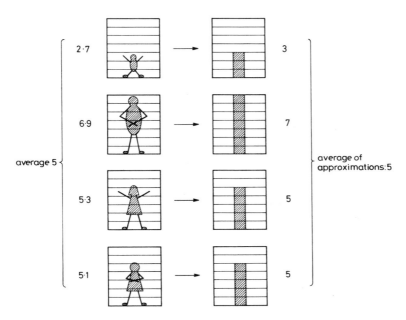

Fig. 5.9. *Why burst processing works*

of designing f.i.r. digital filters from an analogue version is well known and leads to a layout as shown in Fig. 5.10a. To implement such a filter, we simply replace the circuit delays z^{-1} by n-bit shift registers (typically $n = 8$, 10 or 16) and multiply their burst contents (i.e. the number of 'ones') by appropriate constants c_1, c_2, etc. as shown in Fig. 5.10b.

As in arithmetic units there is again a quasianalogue and a purely digital way of realising Fig. 5.10b. In the former case c_1, c_2, etc. are simply the return voltages on b.s.r.s and we use analogue summation. Fig. 5.11 shows such an application for an online Fourier transform unit used in a rather sophisticated frequency analyser for speech (BURFT).[153, 156, 157]

In the digital analyser, the speech signal is represented by its short-time Fourier coefficients. In particular, for voiced speech, a pitch detector determines the speech fundamental period T. N discrete Fourier coefficients are generated from finite samples (each of duration T) of the speech waveform. In the present implementation, $N = 1$. This is the same operation as a spectrum channel vocoder. The speech signal enters an analogue front end where it is amplified and the pitch is extracted (Fig. 5.12). An asynchronous pulse multiplier generates the appropriate sampling clock for the given pitch periods, and this clock drives the transform unit. The resulting coefficients are then used to compute the magnitude of the spectral component.

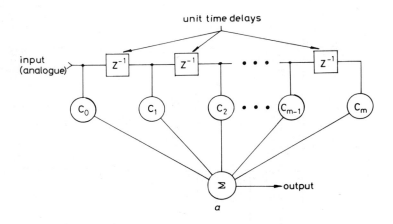

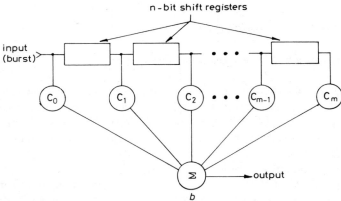

Fig. 5.10.   *Digital filters*
    a General f.i.r. filter
    b Burst version

The problem of convolving the speech signal $Sv(t)$ with $\sin hwt$ and $\cos hwt$ for the hth harmonic is fundamental to the calculation of $|Sv(h)|$. If one divides $Sv(t)$ into $h + 1$ equal segments, only one period of a sine and cosine waveform is required. One merely performs $h + 1$ partial convolutions of each segment with the sine and cosine waves. Summing over these partial convolutions and scaling appropriately, one obtains the coefficients a_h and b_h where

$$a_h = \frac{1}{h+1} \sum_{n=0}^{h} \sum_{k=0}^{d-1} S\left(\frac{nt}{h+1} + \frac{k\Delta t}{h}\right) \sin\frac{2\pi k\Delta t}{T}$$

$$b_h = \frac{1}{h+1} \sum_{n=0}^{h} \sum_{k=0}^{d-1} S\left(\frac{nt}{h+1} + \frac{k\Delta t}{h}\right) \cos\frac{2\pi k\Delta t}{T}$$

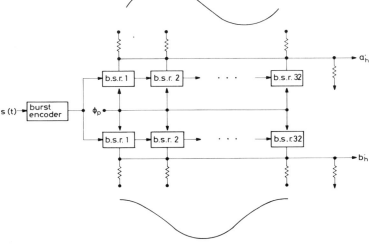

Fig. 5.11. *BURFT Fourier transform unit*

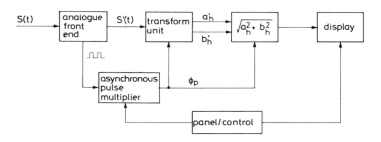

Fig. 5.12. *Speech processor schematic diagram*

This approach was used because the weighting voltages on a row of b.s.r.s can be adjusted to simulate a given waveform. The transform unit shown in Fig. 5.11 consists of two rows of 32 b.s.r.s each. One row is weighted with a sine wave, the other with a cosine wave. It may be shown that by adjusting the sampling rate appropriately, these voltages remain stationary. Each burst-encoded subsection of speech is passed through these two rows. Given the results of the partial convolutions, several operations must be performed to obtain the coefficient. All the partial convolutions for a given fundamental period are added together and then normalised.

This sum is then squared and added to the corresponding sin/cos sum. The magnitude of the spectral line is then produced by taking the square root of this quantity.

These postfiltering operations were implemented in the burst domain to demonstrate another property of this representation. Observing the value of a single burst word, the information is contained in the location of the 1—0 boundary. This positional attribute lends itself to trivial function implementations. By correctly connecting the outputs of a burst register to predetermined inputs of a second register, the receiving register will contain the burst approximation to the function. Given any input distribution, one may reduce the computational error associated with such an approach to any level required. Assuming a fixed word length for input and output, one may increase the number of bits used in the intermediate calculations. A simulation study has shown that the mean square error decreases in an approximate exponential manner with increasing bit length.

In its present form, the transformer samples a speech waveform 32 times to generate the Fourier coefficient for the fundamental frequency. For the coefficient of the hth harmonic, the speech is sampled $h \times 32$ times. In each case, a square window of size T is used. For this reason, the amplitude distortion of the resultant Fourier spectrum can easily be compensated for at the output of the analyser or at the input of the synthesiser.

There are a number of trade-offs in the design. The present version generates one Fourier coefficient every period (T seconds). For different fundamental frequencies, 50 to 500 Fourier coefficients can be produced per second. Twice the output rate can be achieved by grouping the 64 b.s.r.s into four transversal filters instead of two. In this case, the speech waveform is sampled at a rate of $16h/T$ for the hth coefficient. Trading the number of spectral lines for sampling rate can be taken to any appropriate limit. It should be observed that by altering the orthogonal weighting functions on the b.s.r.s, other transformations may be implemented without any significant increase in hardware.

Fig. 5.13 shows an adaptive system (PREDICTORBURST) to calculate the predictor coefficients a_i (only state $i = 2$ is shown explicitly). In view of redundancy of human speech it is well-known that the a_i's vary quite slowly.[83] M.D.A.C. is a multiplying digital-to-analogue converter which sends $a_2 \cdot s_{n-2} [+ a_1 s_{n-1} + a_3 s_{n-3} + \dots]$ to a summing amplifier. The latter forms the error signal. Note the feedback from the error output to the counters storing the a_i.s. Research on such structures has shown that a low-precision number representation which depends on long term averaging such as bursts can be used successfully for the real-time linear prediction of speech. It is quite likely that other applications of such adaptive structures will prove to be feasible.

In order to obtain a purely digital version of a f.i.r. burst filter, we can again revert to a perverted adder (p.a.) design. Here the coefficients $c_1, c_2 \dots$ etc. are supposed to be small integers and their influence is realised by appropriate fan-outs inside a p.a. tree.[158] Using the p.a. tree, a burst filter can be constructed and its performance analysed with standard digital filter theory. Fig. 5.14 shows a general p.a. filter for parallel encoding with block length n. In the design of such a filter, one can easily determine real values for the coefficients c_i given the order of the filter and the desired response, but then they must be 'rounded' to obtain coefficients realisable in a tree. Although the same problem is encountered using weighted-binary representation, use of burst encoding implies a desire to minimise hardware cost, accentuating the problem. Actually, interacting with the problem of finding the coefficients are the questions of filter order and the size of the p.a. tree. Integer programming might be employed to find the best solution, or heuristic methods used to find a suboptimal solution for this complicated trade-off between cost and performance. While this is the difficult step in the design of p.a. filters, it should be recalled that once coefficients are chosen, the characteristics of the filter are found easily.

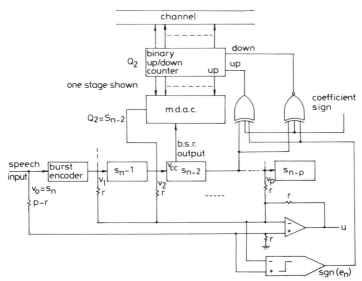

Fig. 5.13.   *PREDICTORBURST unit*

If the output of the filter of Fig. 5.14 is decoded by a b.s.r., we can only guarantee that once every n clock periods (when a full input sample has just entered the filter and the corresponsing output block is located in the b.s.r.) will the b.s.r. output agree with the desired filtered version of the input. Between such samples, because the filter output is not in the burst format, smooth interpolation is not guaranteed. It has been shown that this situation may be remedied by using the same filter with delta block encoding.[150]

Fig. 5.15 shows the practical layout of a Hamming weighted filter.[158] Two clock lines (c.k.s. and c.k.f.) are used because the filter is actually multiplexed. In conjunction with some additional burst filters to suppress the periodic peaks due to sampling at 125 kHz, the response is down 39 dB at 8 kHz from the centre (null) frequency, and down by better than 70 dB at 125 kHz. This is comparable to that which obtains in good communication equipment.

Early research was concerned with the applications of burst techniques in digital radios using comb filters for tuning.[154, 159] The principle is shown (for a.m.) in Fig. 5.16. The idea is that if we sample at the frequency of the desired station, we shall obtain constructive interference of the samples for that station, while other stations will be subject to destructive interference. Here it must, of course, be assumed that during the integration over about 100 cycles the amplitude remains essentially constant. This is the case for speech and a carrier above 1 MHz, a condition always met in real life. In practice it is awkward to operate with a variable frequency clock in order to obtain tuning. Fig. 5.17 shows how this can be avoided by using (delaying) shift registers of variable length. Tuning then corresponds to adjusting the delays so that from b.s.r. to b.s.r. we have exactly one period of the desired station. To extract the a.m. modulation, we simply average the tuned signal, for instance by running it into a counter and watching the overflow.

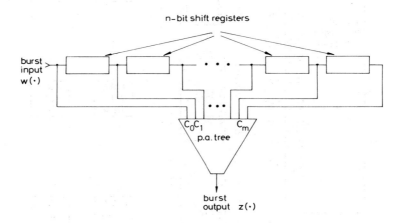

Fig. 5.14.   *General p.a. filter for n-bit blocks*

To demodulate f.m. we must determine the average frequency of the (tuned) antenna signal. This can be done by building a burst-implemented phase-locked-loop (p.l.l.) and watching the average time between zero crossings of its output.[154, 158, 160] The p.l.l.-design in Fig. 5.18 starts, on top, with a zero-crossing detector. Although we indicate a b.s.r. and a comparator in this position, it is clear that we can use a purely digital design. The game is now to count clock-cycles between zero-crossings in an 'instantaneous counter'. If the count is high, we increase the averaging counter; if it is low, we decrease it. Of course, such a feedback system must have some inertia built into it. This is done by delaying the increase/decrease decision in the 'balance register', which has its 1/0 boundary hop up and down, but only large and repeated hops lead to '1' slipping out on top, or

'0' on the bottom. It is the 'slipping out' which modifies the averaging register in the appropriate direction. Clearly the contents of the averaging counter can be used to retrieve the f.m.

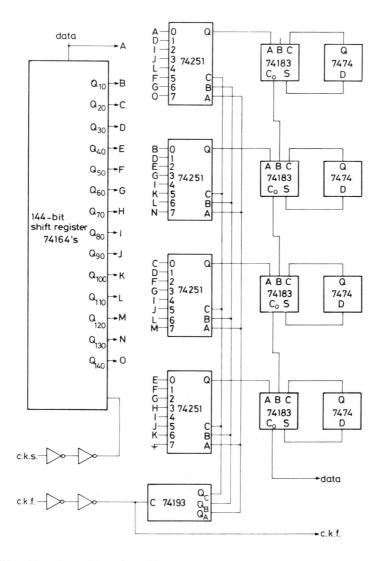

Fig. 5.15. *Hamming weighted burst filter*

Fig. 5.19 shows a circuit diagram for a burst p.l.l.[161] From this figure we see that the input signal zero crossings are used to clock 'ones' through a count register which is cleared following each master clock pulse. If, between clock pulses, 'ones' have succeeded in travelling through the present length of the count register, an error signal is produced which eventually increases the effective register length. If 'ones' have not travelled the present length of the count register between clock pulses, the error signal produced will decrease the effective length of this register.

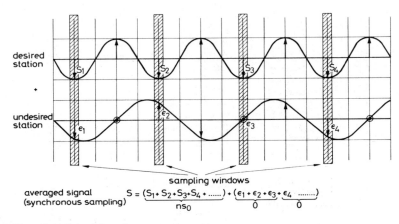

Fig. 5.16.   *Principle of interference tuner*
In both a.m. and f.m. modulation the characteristic amplitude and frequency can be assumed constant over $n = 10-100$ cycles
To tune a station, one arranges to sample at its frequency. (To extract the amplitudes one must actually sample more often!)

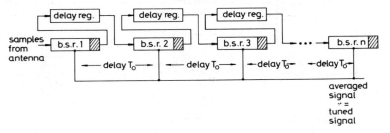

Fig. 5.17.   *Delay register tuner*
To tune a station of period T_0 one changes the length of all delay registers

It can be seen that the number of zeros in the error register, which controls the effective length of the count register, is linearly related to the steady state input frequency. This linearity is an essential requirement when the system is used to demodulate f.m. signals, easily achieved by inverting the contents of the error register and loading it into a b.s.r.

An error delay register is placed inside the feedback loop to provide inertia through its lowpass characteristics. This addition greatly reduces system response to transient spikes and tends to reduce granular noise at the output. Since the clock

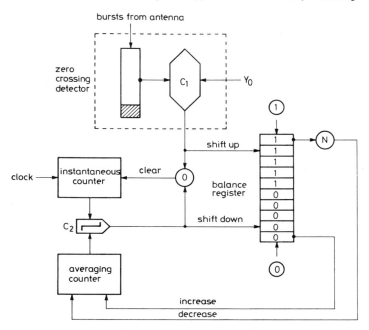

Fig. 5.18. *p.l.l. for f.m. demodulation*

In essence the circuit simply detects whether the next zero crossing occurs (or not) before the instantaneous counter reaches the previous average between zero crossings. The balance register modifies the average.

The width of the window of comparator c_1 corresponds to one pulse of a burst

Comparator c_1 and (simplified) comparator c_2 can be replaced by digital designs

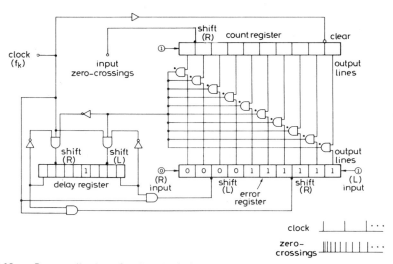

Fig. 5.19. *Burst realisation of a phase locked loop*

and input zero crossings act asynchronously, the count register may not always contain an accurate value of frequency. The maximum error will be minus one bit and will appear at a rate determined by the difference between the clock and input frequencies. The effects of this error alone however will tend to produce no net motion of the 1 contained in the delay register, thus producing an effect in the output. Certainly, the most significant output noise will result from the continual hunting of the system as it tries to track a particular input frequency. The granular noise produced is similar in nature to that found in delta modulators and has a peak amplitude of one bit.

F.M. demodulation via slope detection is another scheme readily adapted to burst processing. This method relies on the fact, that for a constant amplitude f.m. signal, the maximum slope achieved by the carrier is directly proportional to its frequency. Fig. 5.20 depicts the function form of a burst slope detector.[161] From the figure we see that the filtered f.m. signal is sampled at a rate greater than or equal to the Nyquist rate and the encoded burst samples fed into ten bit shift registers.

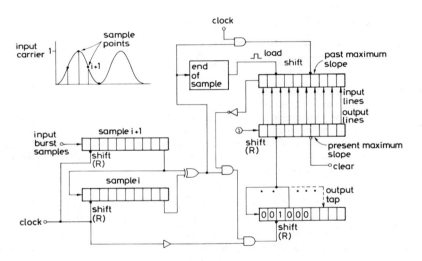

Fig. 5.20. *Slope demodulator implemented with burst hardware*

Adjacent burst samples are then compared on a bit by bit basis by an exclusive-OR which produces, every ten clock periods, a new burst whose value is directly proportional to the magnitude of the difference of the two bursts. If the burst samples are taken at equal time intervals, the value of this difference is proportional to the instantaneous slope of the incoming signal. Equally spaced sample points can be obtained by burst encoding sampled and held values of the input signal, one burst per input sample. The system then proceeds to determine local slope maxima, say over ten slope values, which are periodically loaded into the output b.s.r. to appear as the reconstructed baseband signal.

A third method of f.m. demodulation well suited to burst processing is simply that of counting the zero crossings of the input f.m. signal.[161] It is quite obvious that for a fixed count interval, the number of zero crossings which occurred is directly proportional to frequency. Fig. 5.21 illustrates a basic zero-crossing demodulator using the burst data format. From the figure we see that, for each zero crossing, a 1 is loaded into a 10-bit shift register which is cleared periodically. Before being cleared, the contents of this register are parallel loaded into a buffer register which feeds the output b.s.r. and is clocked at a rate ten times that of the clear pulse. Therefore, the output b.s.r. will contain the instantaneous frequency samples whose variations will be the desired baseband signal.

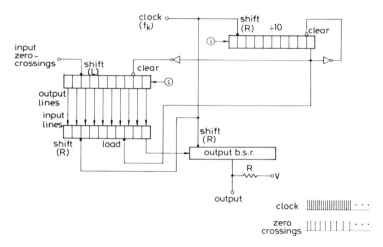

Fig. 5.21. *Zero crossing demodulator using burst techniques*

The area of burst demodulators has proved to be most attractive with various demodulation schemes devised to meet different performance criteria. Apart from analogue comparators and sample-hold modules (if employed), the burst-f.m. receivers employ totally digital processing, delivering the low cost and reliability inherent in digital systems. It appears that in the near future a single b.s.r. could become part of a c.c.d. array containing many such devices, thus further reducing the cost and size of burst systems. With such b.s.r. arrays, 100 slot data representation becomes very practical and permits a tenfold increase in accuracy, thus improving the performance of digital filters and demodulators realised with burst hardware. It is apparent that the use of burst processing in digital f.m. receivers is just in its infancy and further development will no doubt yield more ingenious methods of applying the burst concept to digital recievers.

5.3.2 Picture, indicator – and sequency – processing
In the WALSHSTORE[162] project, the feasibility of storing the 'electronic hologram' of a binary input picture was examined in such a way that the destruction of a large

percentage of the memory would simply correspond to a lack of definitions of the reconstituted input. This objective was attained by using a two-dimensional Walsh-transform. Burst processing was used (in conjunction with weighted binary techniques) because the input came from a c.c.d.-type camera which essentially shifted out bursts. The storage of the transform, (obtained by burst processing) was accomplished in binary, since the storage of a number in nonweighted unary fashion would be prohibitively expensive.

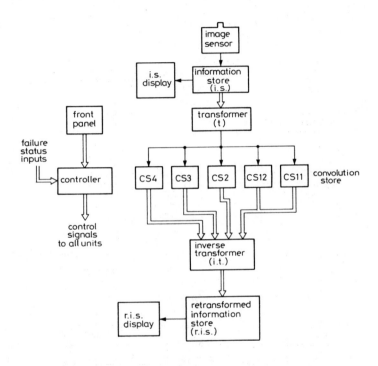

Fig. 5.22.   *Schematic diagram of WALSHSTORE*

Fig. 5.22 shows a schematic diagram of the WALSHSTORE. The information store (i.s.) receives an image from the image sensor and the image is displayed on the i.s. display. A transformer T transforms the contents of i.s. into the convolution store (c.s.). This store is divided into four identical quadrants with the first quadrant duplicated since it stores the most important part of the information. When retrieval of the image is required, the inverse transformer (i.t.) is arranged to transform the contents of c.s. into the retransformed information store (r.i.s.). The division of c.s. into five different parts is arranged so that they may be placed in different physical locations, with the result that their probabilities of failure are independent. The control unit accepts status information from the different quadrants and makes a decision about which one to use should there be duplicates. In addition the control unit supplies all the control signals to the various units during all modes of operation.

It has been demonstrated that t.v. pictures can be encoded, transmitted and

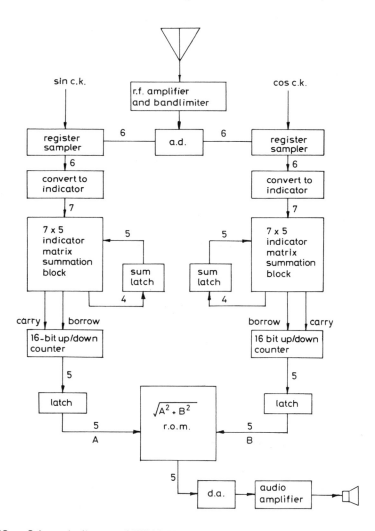

Fig. 5.23. *Schematic diagram of INDIRAD*

decoded with reasonable resolution using burst techniques with only ten levels of intensity.[151] Additionally, there exist image sensors that can provide burst encoded pictures directly; as a result, the image sensor, i.s. and r.i.s. in a WALSHSTORE type system can use burst processing. Since T receives the information from i.s. it is natural for T to use burst processing. Also, it has been shown that the burst T is comparable in complexity to a binary T.[162] WALSHSTORE demonstrates that Walsh transforms can be used in order to achieve a fail-soft storage of pictures and that burst processing is attractive in this application, especially in the front end, i.e. original picture storage and transformer. The use of burst processing in these parts of the system does not increase cost and helps reduce system hardware complexity owing to simple encoder, processors and their controllers.

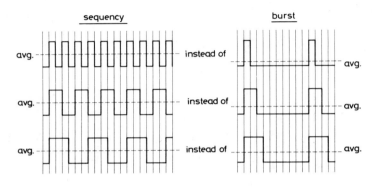

Fig. 5.24.   *Comparison of burst and sequency representation*
Sequency representation is equivalent to quantised pulse-width modulation, but with a variable frame reference (namely the last $0 \rightarrow 1$ or $1 \rightarrow 0$ transition). Successive frames are inverted with respect to each other: the average is always one half of the full swing

Another project INDIRAD[153, 163] used a spatial version of bursts. In a set of seven wires, we energise just one to indicate the numerical value of waveform-samples taken at times in quadrature, i.e. for $\sin t = \pm 1$ and $\cos t = \pm 1$. The advantage of such an 'indicator representation' is that very fast arithmetic becomes trivial. All we have to do is to realise a table-look-up (for sum, product etc.) in hardware. The averaging at the end then smoothes the results. Fig. 5.23 shows the schematic diagram of INDIRAD. The purpose of INDIRAD was to show that indicator processing and quadrature sampling could be used in a digital a.m. receiver. Indicator processing has been shown to be faster than conventional weighted binary processing.[163] The cost of this speed is additional hardware. Beyond about 4 binary bits, or 16 indicator bits, the modest increase in speed does not justify the much greater increase in hardware. However, if there is a need for low precision high speed processing, indicator processing should be considered. Quadrature sampling is thought to be an excellent method of a.m. demodulation. A receiver consisting of an analogue amplifier/bandlimiter, a high order digital filter with 12- to 16-bit precision, and a quadrature sampler is able to achieve a signal/noise ratio exceeding 80 dB.

We have already mentioned that if a 10-slot burst system is replaced by weighted binary, we gain a factor of 2 1/2 in bandwidth. *Sequency Processing*[165] tries to gain bandwidth for the average case (that is, 5 pulses in 10 slots) by abandoning the fixed frame format — this gives a factor 2 on the average, making it nearly the equivalent in bandwidth of weighted binary. The actual format is shown in Fig. 5.24. Again we have a quantised pulse-width modulation, but here we use a variable frame reference (namely the last 0 1 or 1 0 transition). Successive frames are inverted with respect to each other and the average is always one-half of the full swing — this has advantages in carrier wave applications. Although the system updates more frequently than burst processing (as a matter of fact, it has a binary version which updates faster than any other system), we must realise that arithmetic units are more complex, mostly because we have no exact equivalent of the 'windowing property'.

References

1 VON NEUMANN, J.: 'Probabilistic logics and the synthesis of reliable organisms from unreliable components' *in* 'Collected Works', 1963, chap. 6 (Macmillan, New York)
2 PETROVIC, R, and SILJAK, D.: 'Multiplication by means of coincidence' 1962, ACTES Proc. of 3rd Int. Analog Comp. Meeting
3 POPPELBAUM, W. J., AFUSO, C., and ESCH, J. W.: 'Stochastic computing elements and systems' AFIPS FJCC, 1967, **31**, pp. 635–644
4 GAINES, B. R.: 'Stochastic Computing' AFIPS SJCC, 1967, **30**, pp. 149–156
5 AFUSO, C.: Quart. Tech. Prog. Rept., Dept. of Comp. Science, Univ. of Illinois, Urbana, Illinois (starting with January 1964)
6 RIBEIRO, S. T.: 'Comments on pulsed data hybrid computers' *IEEE Trans.*, 1964, EC-13 pp. 640–642
7 POPPELBAUM, W. J.: 'Computer hardware theory' (Macmillan, New York) 1972
8 RING, D.: 'BUM – Bundle processing machine' M.S. Thesis, Rep 353, 1969 Dept. Comp. Science, Univ. of Illinois, Urbana, Illinois
9 CUTLER, J. R.: 'ERGODIC – computing with a combination of stochastic and bundle processing' Rep 630, 1974, Dept. of Comp. Science, Univ. of Illinois, Urbana, Illinois
10 POPPELBAUM, W. J.: 'A practicability program in stochastic processing' Appendix to Proposal to Office Naval Research, March 1974, Dept. of Comp. Science, Univ. of Illinois, Urbana, Illinois
11 POPPELBAUM, W. J.: 'Statistical Processors' *in* 'Advances in Computers', 1976, **14**, pp. 187–230 (Academic Press, New York)
12 AFUSO, C.: 'Analog computation with random pulse sequences' PhD Thesis, Rept. 255, 1968, Dept. of Comp. Science, Univ. of Illinois, Urbana, Illinois
13 RIBEIRO, S. T.: 'Random-Pulse Machines' *IEEE Trans.*, 1967, EC-16, pp. 261–276
14 Proc. of First Int. Symp. on Stochastic Computing and its Applications, Toulouse, France, Nov/Dec 1978
15 GAINES, B. R.: 'Stochastic Computing Systems' *in* 'Advances in Information System Science' (Ed. by J. T. Tou) (Plenum Press, New York) 1969, pp. 37–172
16 GAINES, B. R.: 'Interfaces for stochastic computers with infinite range' Int. Rept., 1976 Dept. of Elec. Eng., Univ. of Essex
17 MILLER, A. J.: 'Digital Stochastic Computation' PhD Thesis, 1976 (Univ. of Aberdeen)
18 MILLER, A. J., BROWN, A. W., and MARS, P.: 'Error reduction techniques in digital stochastic computers' *Trans. IMACS*, 1977, **19**, pp. 51–56
19 CASTANIE, F.: 'Uniformly distributed analog random voltage generator' *Proc. IEEE*, 1978, **66**, (5) pp. 605–606
20 COLDEFY, P. A., CASTANIE, F., and HOFFMAN, J. C.: 'A 1·8 GHz auxiliary source for random reference correlator' Proc. First Int. Symp. on Stochastic Computing, Toulouse, 1978, Paper 2.2, pp. 75–86

21 MASSEN, R.: 'Stochastische Rechentechnik' (C. Hanser Verlag, München) 1977

22 MASSEN, R.: 'A sampling method for correcting the nonlinearities of stochastic converters' Proc. First Int. Symp. on Stochastic Computing, Toulouse, France 1978, Paper 3.2, pp. 129–137

23 GOLOMB, S. W.: 'Shift Register Sequences' (Holden – Day, San Francisco) 1967

24 KORN, G. A.: 'Random Process Simulation and Measurement' (McGraw Hill, New York) 1966

25 TAUSWORTHE, R. C.: 'Random numbers generated by linear recurrence modulo two' *Math. Comput.*, 1965, 9, pp. 201–209

26 PETERSON, W. W., and WELDON, E. J.: 'Error – Correcting Codes' (2nd Ed., MIT Press) 1971

27 MILLER, A. J., and MARS, P.: 'Theory and design of a digital stochastic computer random number generator' *Trans IMACS*, 1977, 19, pp. 198–216

28 MILLER, A. J., BROWN, A. W., and MARS, P.: 'A simple technique for the generation of delayed maximum length binary sequences' *IEEE Trans.*, 1977, EC-26, 8, pp. 808–811

29 SCHWIND, M.: 'On generating and applying a set of independent Bernoulli sequences', Proc. First Int. Symp. on Stochastic Computing, Toulouse, France 1978, paper 2.5, pp. 103–112

30 MILLER, A. J., BROWN, A. W., and MARS, P.: 'Optimum criteria for output interfaces in digital stochastic computers' *Electron. Lett.*, 1975, 11, pp. 326–327

31 BROWN, R. G.: 'Smoothing, Forcasting and Prediction of Discrete Time Series' (Prentice-Hall) 1963

32 MILLER, A. J., BROWN, A. W., and MARS, P.: 'Moving-average output interface for digital stochastic computers' *Electron. Lett.*, 1974, 10, pp. 419–420

33 VINCENT, C. H.: 'Random pulse trains, their measurement and statistical properties' (Peter Peregrinus) 1973

34 MILLER, A. J., BROWN, A. W., and MARS, P.: 'Adaptive logic circuits for digital stochastic computers' *Electron. Lett.*, 1973, 9, pp. 500–502

35 MILLER, A. J., BROWN, A. W., and MARS, P.: 'Study of an output interface for digital stochastic computers' *Int. J. Elect.*, 1974, 37, pp. 637–655

36 FILBERT, D. and GRIMM, H.: 'Mean value computing counters' Proc. First Int. Symp. on Stochastic Computing, Toulouse, France 1978 paper 3.4, pp. 147–156

37 MILLER, A. J., and MARS, P.: 'Optimal estimation of digital stochastic sequences' *Int. J. Sys. Sci.*, 1977, 9, pp. 683–696

38 CASTANIE, F.: 'Generation of random bits with accurate and reproducible properties' *Proc. IEEE*, 1978, 66, pp. 807–809

39 ALBAREDA, A., and CASTANIE, F.: 'Optimisation of the structure of a random number generator with correlated bits', Proc. First Int. Symp. on Stochastic Computing, Toulouse, France 1978, paper 2.4, pp. 89–102

40 CASTANIE, F.: 'Estimation de moments par quantification à référence stochastique' Doctorat-es-Sciences, 1977, Institut National Polytechnique, Toulouse, France

41 HIRSCH, J. J.: 'Contribution à l'étude des conversions numerique – analogique et analogique – numerique au moyen d'une représéntation stochastique de l'information', Thése de Dr-Ingénieur, 1970, Grenoble, France

42 CASTANIE, F.: 'Stochastic conversion – a special case of random quantization' Proc. First Int. Symp. on Stochastic Computing, Toulouse, France 1978, paper 3.1, pp. 113–128

43 HOWARD, B. V.: 'Pulse-Rate Computation: Techniques and Applications' IEE Colloquium on Parallel Processing, April 1976, paper 1

44 MARTIN, J. J.: 'Signal Processing and Computation using Pulse Rate Techniques' *Proc. Inst. Rad. & Electronic Eng.* 1969, 38, pp. 329–344.

45 ESCH, J.W.: 'A display for demonstrating analog computations with random pulse sequences', (POSTCOMP) Rept. 312, 1969, Dept. of Comp. Science, Univ. of Illinois, Urbana, Illinois

46 ESCH, J. W.: 'RASCEL, a programmable analog computer based on a regular array of stochastic computing element logic', PhD Thesis, Rept. 332, 1969, Univ. of Illinois, Urbana, Illinois

47 POPPELBAUM, W. J.: 'What next in computer technology?' *in* 'Advances in Computers', 1968, 9, pp. 1–22, (Academic Press, New York)

48 WO, Y. K.: 'The output display of TRANSFORMATRIX' M.S. Thesis, Rept. 381, 1970, Dept. of Comp. Science, Univ. of Illinois, Urbana, Illinois

49 MARVEL, O. E.: 'TRANSFORMATRIX, an image processor', (Input and stochastic processor sections) PhD Thesis, Rept. 393, 1970, Dept. of Comp. Science, Univ. of Illinois, Urbana, Illinois

50 RYAN, L. D.: 'System and circuit design of TRANSFORMATRIX' (Coefficient processor and output data channel) PhD Thesis, Rept. 435, 1971, Dept. of Comp. Science, Univ. of Illinois, Urbana, Illinois

51 WO, Y. K.: 'A novel stochastic computer based on a set of autonomous processing elements' (APE) PhD Thesis, Rept. 556, 1973, Dept. of Comp. Science. Univ. of Illinois, Urbana, Illinois

52 BROWN, A. W., and MARS, P.: 'Some aspects of the design of a general purpose digital stochastic computer' Proc. First Int. Symp. on Stochastic Computing, Toulouse, France 1978, paper 4.1, pp. 167–191

53 COOMBES, D.: 'SABUMA – safe bundle machine' M.S. Thesis, Rept. 142, 1970, Dept. of Comp. Science, Univ. of Illinois, Urbana, Illinois

54 GAINES, B. R.: 'Stochastic computer thrives on noise' *Electronics*, 1967, 7, pp. 72–79

55 CASTANIE, F.: 'Some new computing elements for stochastic computers' Proc. First Int. Symp. on Stochastic Computing, Toulouse, France 1978 Paper 4.4, pp. 213–222

56 GANDER, J. G.: 'Stochastic generation of sine functions and applications to a new type of PLL' Proc. First Int. Symp. on Stochastic Computing, Toulouse, France 1978, paper 4.5, pp. 225–226

57 GAINES, B. R.: 'Techniques of identification with the stochastic computer' Proc. IFAC Symp. Prague, 1967

58 MARS, P., and McLEAN, H. R.: 'High-speed matrix inversion by stochastic computer', *Electron Lett.*, 1976, 12, pp. 457–459

59 MARS, P., and McLEAN, H. R.: 'Implementation of linear programming using a digital stochastic computer' *Electron Lett.*, 1976, 12, pp. 516–517

60 BEKEY, G. A., and KARPLUS, W. J.: 'Hybrid Computation' (Wiley, 1968)

61 PYNE, I. B.: 'Linear programming on an electronic analogue computer' *AIEE Commun. & Electron.* 1956, 75, pp. 139–143

62 HIRSCH, J. J., and ZIRPHILE, J.: 'Implementation of distributed parameter system models using stochastic representation' Proc. IFAC Cong. Banff, 1971, paper 8.6

63 MARS, P., McINTOSH, F. G., and BAXTER, T.: 'High-speed simulation of discrete dynamic probabilistic systems', *Trans. IMACS*, 1976, 21, pp. 21–38

64 McINTOSH, F. G., and MARS, P.: 'Monte-Carlo simulation using digital stochastic computing techniques', Proc. First Int. Symp. on Stochastic Computing, Toulouse, France 1978, Paper 6.1, pp. 261–276

65 KORN, G. A.: 'Hybrid computer Monte-Carlo techniques' *Simulation*, Oct. 1965, pp. 234–245

66 HANDLER, H.: 'Monte-Carlo solution of partial differential equations' *EE Series Rept.* 16, 1968, Univ. of Arizona, pp. 1–9

67 EDWARDS, K. H., and KHANNA, R. R.: 'Feasibility study of Monte-Carlo modelling techniques for distributed parameter systems' *Proc. IEE*, 1970, 117, (12), pp. 2287–2293

68 HERAULT, J.: 'Nervous system considered as a stochastic processor: application to information processing' Proc. First Int. Symp. on Stochastic Computing, Toulouse, France 1978, paper 6.2, pp. 277–286

69 RETTER, M. L.: 'Space – time structure in tensor transforms' Proc. First Int. Symp. on Stochastic Computing, Toulouse, France 1978, Paper 8.4, pp. 411–420

70 CUTLER, J. R.: 'Molecular Stochastics: a study of direct production of stochastic sequences from transducers' PhD Thesis, Rept. 723, 1975, Dept. of Comp. Science, Univ. of Illinois, Urbana, Illinois

71 MASSEN, R.: 'Stochastic fluidic computing systems' Proc. Fifth Cranfield Fluidics Conf., 1972, pp. 45–55

72 MASSEN, R.: 'Stochastic and other time – summation fluidic digital to analog converters' Proc. Sixth Cranfield Fluidics Conf., 1974 Cambridge, UK, E2, pp.15–28

73 TUMFART, S.: 'New instruments use probabilistic principles' *Electronics,* 1975, **48**, July, pp. 86–91

74 BÜSCH, G., FILBERT, D., and LIEBMANN, G.: 'A modular system for stochastic ergodic measuring technique', Proc. First Int. Symp. on Stochastic Computing, Toulouse, France 1978, paper 8.2, pp. 379–391

75 GANDER, J. G.: 'A simple stochastic implementation of the linear prediction filter' Proc. First Int. Symp. on Stochastic Computing, Toulouse, France 1978, paper 4.2, pp. 193–201

76 VERNIERES, F., CASTANIE, F., and HOFFMAN, J. C.: 'Stochastic one bit wide-band spectrum analyser' First Int. Symp. on Stochastic Computing, Toulouse, France 1978, Paper 8.3, pp. 393–409

77 KUSCHE, P., MASSEN, R., SEIBT, A., and VIAENE, O.: 'An electricity counter with nonlinear stochastic encoding' Proc. First Int. Symp. on Stochastic Computing, Toulouse, France 1978, paper 5.1, pp. 227–236

78 CORRADETTI, M., and OLIVA, I.: 'MOS A/D and D/A convertor circuits based on the stochastic principle', Proc. Fifth Int. Conf., Microelectronics, Munich, Nov. 1972

79 CUTLER, J. R., and FICKE, D.: 'A stochastic control system' Rept. 752, 1975, Dept. of Comp. Science, Univ. of Illinois, Urbana, Illinois

80 MIRAMBET, P., CASTANIE, F., and HOFFMAN, J. C.: 'Industrial application of stochastic computing to turbine regulation' Proc. First Int. Symp. of Stochastic Computing, Toulouse, France 1978 paper 5.3, pp. 249–260

81 CRABERE, P., and VERDIER, R.: 'Feasibility study and implementation of the roll-axis fly-by-wire law, applicable to a supersonic transport type aircraft using stochastic computing' Proc. First Int. Symp. on Stochastic Computing, Toulouse, France 1978 Paper 5.2, pp. 237–248

82 FERRATE, G. A., POIGJANER, L., and AGULLO, J.: 'Introduction to multichannel stochastic computation and control' Proc. IFAC Congress, Warsaw 1969

83 PITT, D. A.: 'Bandwidth compression of speech using linear prediction and burst processing' Proc. First Int. Symp. on Stochastic Computing, Toulouse, France 1978, Paper 1.4, pp. 55–63

84 PITT, D. A., POPPELBAUM, W. J., and XYDES, C. J.: 'OPTOBUNDLE – A Unique Fibre Optic Multiplier' Rept. 882, 1977 Dept. of Comp. Science, Univ. of Illinois, Urbana, Illinois

85 GAINES, B. R.: 'Trends in stochastic computing' IEE Colloquium on Parallel Processing Paper 4, 1976

86 GAINES, B. R.: 'The role of randomness in cybernetic systems' Proc. Conf. on Recent Topics in Cybernetics, Cybernetics Society, London, Sept. 1974

87 GAINES, B. R.: 'The role of randomness in systems theory' Int. Rept. EES-MMS-RAN-76, Dept. of Elect. Eng. Sci., Univ. of Essex, 1974

88 GAINES, B. R.: 'Memory minimisation in control with stochastic automata' *Electron. Lett.,* 1971, **7**, pp. 710–711

89 BUSH, R. R., and MOSTELLER, F.: 'Stochastic models for learning' (Wiley, New York) 1958

90 ATKINSON, R. C., BOWER G. H., and CROTHERS, E. J.: 'An introduction to mathematical learning Theory' (Wiley, New York) 1965

91 IOSIFESCU, M., and THEODORESCU, R.: 'Random processes and learning' (Springer-Verlag) 1969

92 NORMAN, M. F.: 'Markov processes and learning models' (Academic Press, New York) 1972

93 NARENDRA, K. S., and THATHACHAR, M. A. L.: 'Learning automata – A survey' *IEEE Trans.*, 1974, SMC-4, pp. 323–334

94 Special Volume on Learning Automata, *J. Cybern. & Inf. Sci.*, 1, (2), 1977

95 NARENDRA, K. S., and THATHACHAR, M. A. L.: 'On the behaviour of a learning automaton in a changing environment with application to telephone traffic routing' S & IS Report No. 7803, Yale University, October, 1978

96 SRIKANTA KUMAR, P. R., and NARENDRA, K. S.: 'Learning algorithm models for routing in telephone networks' S & IS Report No. 7903, Yale University, May, 1979

97 TSETLIN, M. L.: 'On the behaviour of finite automata in random media' *Automat i Telemekh*, 1961, 22, pp. 1345–1354

98 TSETLIN, M. L.: 'Automaton theory and modelling of biological systems' (Academic Press, New York) 1973

99 KRYLOV, V. U.: 'On one stochastic automaton which is asymptotically optimal in a random medium' *Automation and Remote Control*, 1963, 24, pp. 1114–1116

100 KRINSKII, V. I.: 'An asymptotically optimal automaton with exponential rate of convergence' *Bio Physics*, 1964, 9, (4), pp. 484–487

101 PONOMAREV, V. A.: 'A construction of an automaton which is asymptotically optimal in a stationary random medium' *Bio. Physics*, 1964, 9, (1), pp. 104–110

102 MACKIE, N. J., and MARS, P.: 'Stochastic automata in non-stationary environments' Proc. First Int. Symp. on Stochastic Computing, Toulouse, France 1978, paper 7.3, pp. 321–344

103 LOUI, M. C., and NARENDRA, K. S.: 'Comparison of learning automata operating in non-stationary environments' Tech. Rept. CT-65, Becton Centre, Yale Univ., 1975

104 VARSHAVSKII, V., and VORONTSOVA, I.: 'On the behaviour of stochastic automata with variable structure' *Automat i Telemekh*, 1963, 24, pp. 353–360

105 McLAREN, R. W.: 'A stochastic automaton model for synthesis of learning systems' *IEEE Trans.*, 1966 SSC-2, pp. 109–114

106 LAKSHMIVARAHAN, S., and THATHACHAR, M. A. L.: 'Optimal non-linear reinforcement schemes for stochastic automata', *Info. sci.*, 1973, pp. 78–103

107 COUTTS, M. J., and MARS, P.: 'Study of a modified–estimating automaton in stationary environments' *Electron. Lett.*, 1978, 14, (13) pp. 404–406

108 NEVILLE, R. G., NICOL, C. R., and MARS, P.: 'Design of stochastic learning automata using adaptive digital logic elements' *Electron. Lett.*, 1978 14, pp. 324–326

109 NEVILLE, R. G., NICOL, C. R., and MARS, P.: 'Synthesis of stochastic learning automata' *Electron. Lett.*, 1978, 14, pp. 206–208

110 NEVILLE, R. G., and MARS, P.: 'Design of nonlinear stochastic learning automata' *Electron. Lett.*, 1978, 14, pp. 396–397

111 VISWANATHAN, R., and NARENDRA, K. S.: 'Application of stochastic automata models to learning systems with multimodal performance criteria' Becton Centre, Yale Univ., Tech. Rept., CT-40, June 1971

112 NARENDRA, K. S., and VISWANATHAN, R.: 'A two-level system for periodic random environments' *IEEE Trans.*, 1972, SMC-2, pp. 258–289

113 NEVILLE, R. G., and MARS, P.: 'Hardware design for a hierarchical structure stochastic learning automaton' *J. of Cybern. & Inf. Sci.*, 1979, 2, (1), pp. 30–35

114 NEVILLE, R. G., and MARS, P.: 'Hardware synthesis of stochastic learning automata' Proc. First Int. Symp. on Stochastic Computing, Toulouse, France 1978 Paper 7.4, pp. 345–365

115 McMURTRY, G. J., and FU, K. S.: 'A variable structure automaton used as a multimodal search technique, *IEEE Trans.*, 1966, AC-11, pp. 379–387

116 SHAPIRO, I. J., and NARENDRA, K. S.: 'Use of stochastic automata for parameter self-optimisation with multimodal performance criteria' *IEEE Trans.*, 1969, **SSC-5**, pp. 352–360

117 JARVIS, R. A.: 'Adaptive global search by the process of competitive evolution' *IEEE Trans.*, 1975, **SMC-5**, pp. 297–311

118 GINSBURG, S. L., and TSETLIN, M. L.: 'Some examples of simulation of the collective behaviour of automata' *Problemi Peredachii Informatsii*, 1965, **1**, (2), pp. 54–62

119 STEFANYUK, N. L., and TSETLIN, M. L.: 'Power regulation in a group of radio stations' *Problemi Peredachii Informatsii*, 1967, **3**, (4), pp. 49–57

120 VARSHAVSKII, V., MELESHINA, M. L., and TSETLIN, M. L.: 'Priority organization in queueing systems using a model of collective behaviour' *Problemi Peredachii Informatsii*, 1968, **4**, (1), pp. 73–76

121 RADYUK, L. E., and TERPUGOV, A. F.: 'Effectiveness of applying automata with linear tactic in signal detection systems' *Automation and Remote Control*, 1971, **32**, pp. 99–107

122 LI, T. J., and FU, K. S.: 'Automata games, stochastic automata and formal languages' Tech. Rept. TR-EE69-1 Purdue Univ. Lafayette, 1969

123 GLORIOSO, R. M., GRUENEICH, G. R., and McELROY, D.: 'Adaptive routing in a large communications network' Proc. IEEE Symp. on Adaptive Processes, 1970, IEEE Publication 70C 58-AC, pp. 5.1–5.4.

124 GLORIOSO, R. M., GRUENEICH, G. R., and DUNN, J. C.: 'Self-organisation and adaptive routing for communication networks' Proc. EASCON 1969, IEEE Publication 69C 31-AES

125 NARENDRA, K. S., MASON, L. G., and TRIPATHI, S. S.: 'Application of learning automata to telephone traffic routing problems' Tech. Rept. CT-60, Becton Centre, Yale Univ. 1974

126 NARENDRA, K. S., and WRIGHT, E. A.: 'Application of learning automata to telephone traffic routing problems' Tech. Rept. CT-69, Becton Centre, Yale Univ. 1976

127 GLORIOSO, R. M., and COLON-OSORIO, F. C.: 'Cybernetic control of computer networks' Fifth Annual Modelling and Simulation Conf., Pittsburgh, 1974

128 COLON-OSORIO, F.: 'Scheduling in multiple-processor systems with the aid of stochastic automata' Ph.D. Dissertation, Univ. of Mass., 1977

129 ASAI, K., and KITAJIMA, S.: 'A method for optimizing control of multimodal systems using fuzzy automata' *Inf. Sc.*, 1971, **3**, pp. 343–353

130 EL FATTAH, Y. M., and NAJIM, K.: 'Use of a learning automaton in static control of a phosphate drying furnace' Proc. of 5th IFAC/IFIP Int. Conf., Hague, 1977

131 NARENDRA, K. S., WRIGHT, E. A., and MASON, L. G.: 'Application of learning automata to telephone traffic routing and control, *IEEE Trans.*, 1977, **SMC-7**, pp. 785–792

132 NARENDRA, K. S., and McKENNA, D. M.: 'Simulation study of telephone traffic routing using learning algorithms – Part I' S & IS Rept. No. 7806, December, 1978

133 NARENDRA, K. S., MARS, P., and CHRYSTALL, M. S.: 'Simulation study of telephone traffic routing using learning algorithms – Part II' S & IS Rept. No. 7907, Yale Univ. October, 1979

134 MARS, P., and CHRYSTALL, M. S.: 'Real-time telephone traffic simulation using learning automata routing' S & IS Rept. No. 7909, Yale Univ. November 1979

135 HILLS, M. T.: 'Telecommunications switching principles' (MIT Press, Cambridge) 1979

136 BOEHM, B. W., and MOBLEY, R. L.: 'Adaptive routing techniques for distributed communication systems, *IEEE Trans.*, 1969, **COM-17**, pp. 340–349

137 FULTZ, G. L., and KLEINROCK, L.: 'Adaptive routing for store-and-forward computer communication networks' Proc. 1971, Int. Conf. Comm., Montr., Canada, June 1971. pp. 39.1–39.8

138 PROSSER, R. T.: 'Routing procedures in communication networks – Part I' *IRE Trans.*, 1962, **CS-10**, pp. 322–329

139 FULTZ, G. L.: 'Adaptive routing techniques for message switching computer communication network' School of Eng. and App. Sci., Univ. of California, UCLA-ENG-7252, June, 1972

140 FRANK, H., and FRISCH, I. T.: 'Communication, transmission and transportation networks', Addison Wesley, 1971

141 HANSLER, E.: 'An experimental heuristic procedure to optimize a telecommunication network under nonlinear cost function' Proc. 7th Ann. Princeton Conf. Inf. Sci. and Systems, 1973. pp. 130–134

142 CHOU, W., and FRANK, H.: 'Routing strategies for computer network design' Proc. Symp. Computer Commun., Poly. Inst. of Brooklyn, 1972. pp. 301–309

143 FRATTA, L., GERLA, M., and KLEINROCK, L.: 'The flow deviation method: An approach to store and forward communication network design' *Networks*, 1973, **3**, (3), pp. 97–133

144 SCHWARTZ, M., and CHEUNG, C. K.: 'Alternate routing in computer-communication networks' Proc. 7th Hawaii Int. Conf. System Sciences, Jan. 1974. pp. 67–69

145 BEIZER, B.: 'A comparative Evaluation of three message switched networks' Data System Analysts, Rep. Im-5053, Oct. 1973

146 BROWN, C. W., and SCHWARTZ, M.: 'Adaptive routing in centralized computer-communication networks' Proc. IEEE Int. Conf. on Communications, San Francisco, June, 1975. pp. (47–12)–(47–16)

147 RUDIN, H.: 'On routing and "Delta-Routing": A taxonomy and performance comparison of techniques for packet-switched networks' *IEEE Trans.*, 1976, **COM-24**, (1), pp. 43–59

148 NEVILLE, R. G., CHRYSTALL, M. S., and MARS, P.: 'Application of hierarchical structure stochastic learning automaton' S & IS Rept. No. 7906, Yale Univ. September, 1979

149 POPPELBAUM, W. J.: 'Burst processing: a deterministic counterpart to stochastic computing' Proc. First Int. Symp. on Stochastic Computing, Toulouse, France. 1978, Paper 1.1, pp. 1–30

150 TAYLOR, G. L.: 'An analysis of burst methods and transmission properties' Rept. 770, 1975, Dept. of Comp. Sci. Univ. of Illinois, Urbana, Illinois

151 WOLFF, M: 'Transmission of analogue signals using burst techniques' Rept. 838, 1977, Dept. of Comp. Sci. Univ. of Illinois, Urbana, Illinois

152 WOLFF, M.: 'Digital encoding and decoding using burst techniques' Proc. First Int. Symp. on Stochastic Computing, Toulouse, France 1978, Paper 1.2, pp. 31–44

153 POPPELBAUM, W. J.: 'Application of stochastic and burst processing to communication and computing systems' Proposal to ONR, 1976, Dept. of Comp. Sci. Univ. of Illinois, Urbana, Illinois

154 POPPELBAUM, W. J.: Proposal to ONR entitled 'Application of stochastic and burst processing to communication and computing systems, 1975, Dept. of Comp. Sci. Univ. of Illinois, Urbana, Illinois

155 TIETZ, L. C.: 'Burstlogic: design and analysis of logic circuitry to perform arithmetic on data in the burst format' Rept. 895, 1977, Dept. of Comp. Sci. Univ. of Illinois, Urbana, Illinois

156 XYDES, C. J.: 'Application of burst processing to the spectral decomposition of speech' Rept. 870, 1977, Dept. of Comp. Sci. Univ. of Illinois, Urbana, Illinois

157 XYDES, C. J.: 'The decomposition of speech using burst processing' Proc. First Int. Symp. on Stochastic Computing, Toulouse, France, 1978. Paper 1.3, pp. 45–54

158 WELLS, D. K.: 'Digital filtering using burst processing techniques' Rept. 871, 1977, Dept. of Comp. Sci. Univ. of Illinois, Urbana, Illinois

159 MOHAN, P. L., BRACHA, E., and LIU, J. W. S.: 'Performance evaluation of the digital f.m. receiver' Rept. 757, 1975, Dept. of Comp. Sci. Univ. of Illinois, Urbana, Illinois

160 ROBINSON, C. M.: 'Burstlock: a digital phase-locked loop using burst techniques' Rept. 872, 1977, Dept. of Comp. Sci. Univ. of Illinois, Urbana, Illinois

161 MOHAN, P. L.: 'The application of burst processing to digital FM receivers' Rept. 780, 1976, Dept. of Comp. Sci. Univ. of Illinois, Urbana, Illinois

162 BRACHA, E.: 'WALSHSTORE: The application of burst processing to fail-soft storage systems using Walsh transforms' Rept. 878, 1977, Dept. of Comp. Sci. Univ. of Illinois, Urbana, Illinois

163 GOSTIN, G. B.: 'A digital receiver for amplitude modulated signals using indicator processing' Rept. 923, 1978, Dept. of Comp. Sci. Univ. of Illinois, Urbana, Illinois

164 BRACHA, E.: 'BURSTCALC (A BURST CALCulator), Rept. 769, Dept. of Comp. Sci. Univ. of Illinois

165 POPPELBAUM, W. J.: 'Application of stochastic and burst processing to communication and computing systems' Proposal to ONR, 1977, Dept. of Comp. Sci. Univ. of Illinois, Urbana, Illinois, Urbana, Illinois

166 PLEVA, R. M.: 'A microprocessor-controlled interface for burst processing' Rept. 812, 1976, Dept. of Comp. Sci. Univ. of Illinois, Urbana, Illinois

167 MA, J. H. C.: 'Block sum register and burst arithmetic' Rept. 890, 1977, Dept. of Comp. Sci. Univ. of Illinois, Urbana, Illinois

Index